MAGIC: SYMBOLS AND TEXTS OF MAGIC

Luis Carlos Molina Acevedo

Title: Magic: Symbols and Texts of Magic

First edition

Copyright ©2016 Luis Carlos Molina Acevedo

© Texts: Luis Carlos Molina Acevedo

Author: Luis Carlos Molina Acevedo

Contact: lcmolinaa@yahoo.es

http://lcmolinaa.blogspot.com

Cover Design: Luis Carlos Molina Acevedo

Review of Stile: Luis Carlos Molina Acevedo

All Rights Reserved

ISBN-13: 978-1532807053

ISBN-10: 1532807058

About the Author

Luis Carlos Molina Acevedo was born in Fredonia, Colombia. He is Social Communicator of the University of Antioquia, and Masters in Linguistics from the same university. The author has published the next books on online bookstores:

I Want to Fly, From Don Juan to Sexual Vampirism, The Imaginary of Exaggeration, The Clavicle of Dreams, For Writers by Writers, The Modern Concept of Communication, Is There Anybody Out The Wall?, Dr. House Syndrome, Zombie Factor, and Magic: Symbols and Texts of Magic.

Quiero Volar, El Alfarero de Cuentos, Virtuales Sensaciones, El Abogado del Presidente, Guayacán Rojo Sangre, Territorios de Muerte, Años de Langosta, El Confesor, El Orbe Llamador, Oscares al Desnudo, Diez Cortos Animados, La Fortaleza, Tribunal Inapelable, Operación Ameba, Territorios de la Muerte, La Edad de la Langosta, Del Donjuanismo al Vampirismo Sexual, Imaginaria de la Exageración, La Clavícula de los Sueños, Quince Escritores Colombianos, De Escritores para Escritores, El Moderno Concepto de Comunicación, Sociosemántica de la Amistad, Magia: Símbolos y Textos de la Magia, ¿Hay Alguien Afuera del Muro?, Síndrome Dr. House, y Factor Zombi.

Contend

Presentation

This text is the result of research about the culture of south-western in Antioquia, Colombia. It begins in 1986. It was based on the oral tradition. It was supplemented by consulting other sources: files and printed texts.

For this specific theme about magical discourse, they were done specific tasks of research between 1990 and 1991. The field work was conducted in the municipality of Fredonia, Colombia. It was supplemented with a review of literature about magic. The one considered here, it is primarily a theoretical reflection. It aims to provide a linguistic and semiotic methodology for the study of the magical phenomenon. The result was presented as Thesis for the degree of Master in Linguistics at the University of Antioquia, in 1991. The thesis was titled "Text and Context of magic".

The reader will find here, the data obtained from fieldwork. This latter, it marks and defines the relevance of the theoretical elements adopted for the explanation and analysis of the magical phenomenon. At all times, the magical praxis determined what theories should be incorporated into this analysis, always in an effort to clarify the theme. These theories allowed showing the universal scope of magic. His discursive and symbolic features are repeated in time and space.

In this version with book form, all statistical tabulation of fieldwork was excluded. In this, they were done 173 recorded interviews, 87 to women and 86 to men. These were based on a questionnaire with 69 questions; some of closed type and others open type. With that, I sought to obtain quantitative and qualitative data. Because of the characteristics of this study, priority was given to the latter, that is, the qualitative data. The questions explored different aspects of the four manifestations of magic in Fredonia: curers [healers or quacks, men who heal with secrets of the nature: **healer-magician**] (12 questions), orisoners [prayers or supplicants, men who heal with orisons: **prayer-magician**] (5 questions), herbalers [herbalists, men who heal with herbals: **herbalist-magician**] (6 questions), helpers [assisted-men, men who heal with assisting of spirits: **helper-magician**] (18 questions). With quantitative data, it was obtained a weighting index which is equivalent to 58.70 percent. This result says that for every 100 inhabitants in the municipality of Fredonia, 58.70 believe in magic and his four manifestations. This index is something amazing. Fredonia is close to Medellin, capital of Antioquia department. It has also good medical services. But, for every person to go to the doctor there is also one using magic resources. In the case of helper-magician, the weighting index is the highest with 67 percent. That is, for every hundred inhabitants, 67 practice or make use of magical acts aided by supernatural beings.

In this version with book form, some theoretical considerations were also excluded by being too specialized. I expect that the one presented here, it is sufficient to understand the magic in its various constituent elements.

In a work of this nature, there are many people who should be grateful. I want especially to refer to Gabriel Jaramillo Echeverri. He always encouraged me to keep looking the amazing and wonderful culture of Antioquia. The same I should say about Professor Olimpo Suarez for his methodological suggestions. Teachers of the Master in Linguistics were also important for their teachings. I thank all the informants for their selfless collaboration. I apologized for not to mention all those who in one way or another contributed to the completion of this study.

PART I: theoretical framework

This text is a modification of the thesis "Text and Context of magic" presented to obtain the title of:

Master in Linguistics

University of Antioquia

Medellin - Colombia

1991

The modification is aimed at giving a book form to the content mentioned.

Luis Carlos Molina Acevedo

Roadmap

In the present study about magic, it is interested mainly about linguistic and semiotic aspects. In the analysis, we start from the following definition about magic:

Magic is objectified language and symbols to work with power over the world, by a magician to the particular interest of a user into a community, whose interaction constitutes a system.

This definition can be broken down into its constituent elements as follows:

Magic

Language

Symbol

Object

Act

Power

World

Magician

Particular interest

User

Community

Interaction

System

The study, of these elements, demand specific and analytical tools, as follows:

1. The linguistics and semiotics to study the expression of magic and its forms of operating.

2. The Information Theory and its science involved (Dianetics and bioenergetics) to show how magic effectiveness is reached.

3. The Theory of Post-modernity to account for the scenic characteristics of the magical ritual. The aesthetic theatre of magic is critical to production of emotional reactions, leading to disease or healing by magic action.

4. The Theory of Speech Act and Philosophy of Symbolic Forms, to show how the magic manages to operate with power over the world and its creatures.

5. The Theory of Discourse, to establish the structure of magical discourse.

6. The Theory of Symbols, to establish the symbolic structure of magic.

Basic Concepts

Linguistics and semiotics provide elements for a new look of magic. We have not thought enough about the implications of this social practice in our environment. Explanations have been provided from outside. Levi-Strauss in 1950 in his "Structural Anthropology" raised a vision of magic in Brazil. Artaud with his "Trip to the Taraumaras" makes a literary approach to magic practices from peyote in Mexico. Mircea Eliade from his history of religions has characterized the symbolic scope of magic. Jung makes psychoanalysis of folklore, including magic. One could continue mentioning names to conclude: References to this activity in Latin America, they have been external. The few inner visions of the phenomenon have remained in the descriptive and anecdotal like the one done by Carlos Castaneda in Mexico.

The existence of magic goes back to the very origins of man as a thinking being. If their different manifestations are observed, through history, its operation has a basic assumption. It can be formulated as: magic is objectified language by the imagination to operate with power over the world. This assumption authorized us to speak about two discourses. One is behind the magic. It is its essence itself. The other legitimates it. Thus, the key elements in approaching the study of magical discourse are: **magic**, **language**, **symbol** (like objectified language), **imagination**, **action** as operate, the concept of **power** in magic, and the notion of the **world**. These seven concepts will be the focus of this study.

Magic as human action requires a discourse. It nourishes and designates. One of the main legitimizing discourses of magic is the occultism. "It must understand that the theories on which is based occultism are linked to a way of thinking of which we went away by Descartes and discursive thought" (Tondriau, 1966: 13). In the occultism (the method), it is experimental, intuitive and, above all, analog. Jorge Luis Borges wrote in this regard: "Magic is the crowning or the nightmare of fortuity, not its contradiction. The miracle is not more common in the universe than in the one of astronomers. Its world, it is governed by natural laws and by other imaginaries" (Tondriau, 1966: 14). That is, the magic, like all other occult disciplines, refers to a way of thinking. It is opposed to scientific thought, but no less coherent.

In addiction, the magical phenomenon demands also an approach of both, in its psychic manifestation and in its social configuration. At first, it can be considered as "the psychic process to impose the hallucinatory presence of an external object nonexistent" (Molina Acevedo, 1990: 48). But, to this conception, it should make some clarifications in light of new research. A characterization of magic should consider: it is a **social action**, use **symbols**, use **language**, has fixed **texts**, and needs **power**, among other things. Some authors, from different points of

6

view, have tried to clarify the practice of magic. Thus, for Castiglioni (1947: 315-320), the magic is characterized by:

1. It requires preparation of the atmosphere through actions, reminiscence of symbolic and magical rites.

2. It is originated by phenomena that occur in the unconscious.

3. It derives its actions from a mechanical automatism.

4. It needs an intermediary, the magician.

5. It leaves open the possibility for enemy spirits who can hinder action.

6. It needs the incorporating a desire with objective form.

This view offers a wide panorama and not one narrow like the one by Levi-Strauss, based only on the belief:

"There is no reason, therefore, to doubt the efficacy of certain magical practices. But at the same time, it is noted that the effectiveness of magic involves belief in magic, and it comes in three complementary aspects: first, the belief by the sorcerer in the effectiveness of his techniques; then, the one of the ill cared by him or the one of the victim pursued by him, in the power of the sorcerer himself; finally, trust and demands from collective opinion, which is at every moment a kind of gravitational field within which are defined and are situated relations between the sorcerer and those he has bewitched" (1969: 152).

Belief is not enough to explain the complexity of magic. The model of the world about a community, in turn, it is disintegrated into individual models. Through social interaction, they are changed or adjusted. So, magic should be seen as a social system fairly complex.

The Magic System

A great way to understand the operation of magic, It is seeing it as a system. To set it, it is interesting to follow the introductory lines by Ignacio Gomez de Liano in Bruno (1982: 224). He says, the magic work undertaken by Giordano Bruno, includes technical-writing-body-ghost. With some modifications, one can set to these as the elements of the magical system in the light of recent concepts as well:

Technical — Language — Magician — Numen → User + Group.

Activation of this system, it is directed to reinvent and rediscover the world of the patient, also of his group through the re-description and reinterpretation.

The technique within the magic system is composed of a number of procedures and requirements to use the magical instruments. In the jargon of this discourse, they form the 'magic kitchen', within which the components are also counted for the preparation of magical compounds. It is interesting to note how from a system such as science, this technique is considered a mere artifice.

Language acts as a means for operating power over the world. But in turn, it is the means to link supernatural forces to magical acting. It is characterized by its petrification or fixity in 'magic formula', 'recipes', 'secret' and 'prayers', where it is kept unchanged. It is based on the expression of start language, the one of the creation of the world sprouting from breath of God. The tradition considers, if it changes, it will lose its effectiveness.

"In its first form, as was given by God to men, the language was a sign absolutely certain and transparent of things, because it seemed to them. The names were deposited on what they designated as the force is written in the body of the lion, royalty in the eyes of the eagle and as the influence of the planets is marked on the forehead of men: by the shape of the similarity" (Foucault, 1984: 44).

The magician supports the technical and language. He receives strength of numen to act as a liaison between the user and the group. It is the physical part of the system where magic remains and is transformed. It is the microcosm where the macrocosm is repeated. For Foucault: "If it interrogates the knowledge of the XVI century at level archaeological, -that is, in what has done it possibly-, they appear the relations between the macrocosm and microcosm as a simple surface effect" (1984: 40). It should be added, before the magic come to shape the episteme of resemblance, this was and still is the way of thinking in the magical action wherever it is manifested.

But beyond the presence, the magician is the one by the supernatural power. He can manage and control it. That makes him different from other men. Not all members of the group can become magicians in the sense professional of the term or in social division of labour. This aspect of power rescued by the sociological theory of magic, it is fundamental in the analysis of magical discourse. "A purely demonological theory of magic would not be enough to explain the power of magicians, virtue of words, the effectiveness of the gestures, the power of the gaze, of the intention, of the fascination of death" (Cazeneuve, 1971: 139).

The numen is the extra strength, needed by the magician to operate. It is based on the principle: everything is up, in the macrocosm, is in the microcosm, in man. This principle is the essence of power to act on the world. It is the process of linking macro forces through certain rituals. The numen has received many names, 'aura', 'astral' 'plasmic body', 'Biofield', 'animal magnetism' among others. To this principle comes up the magic to explain its procedures, effectiveness and raison to be.

The constant features of the numinous symbolism are those of the unusual, abnormal, and exceptional. That explains why communities choose, as magician, persons unusual. The preferred ones are the physical limited or low harmonics physical features. To Cazeneuve (1971: 137-146), the numinous powers handled by the magician can specify that margin of unpredictability. Human destiny is not reduced to rules. When it always manifests in one sense, good or bad luck, the common man considers it as revealing a strange force, good or bad luck. It gets to recognition by man of something supernatural. His fate, it is no longer dependent on the natural action.

The user is the excuse or reason to be for activation of magic as a social system. For it, it is put into play the art, language and objectified relationship with the numen. The user is who exposes to the magician the problems to solve by magic procedures. In addition, the magician can act, in turn, as user for own affairs. He does also works aimed at obtaining personal results. Any request is a social dysfunction. The individual claims to be aligned back to the social system through the magician. Dysfunctions in magic are alterations in social transactions of the individual.

The user inputs the social character to magic. Thus, the magician does not have his art as a profession or means of subsistence, the fact of being possessor of power, he is undertaken to provide the required assistance. Something quite different happens with no-magician, who boasts of being one by possessing knowledge of the recipe and have access to traditional knowledge, and yet lacks the power. "There, where it is disappeared the disturbing force of disturbing object of the magic, there remains only the technique and skill" (Cazeneuve, 1971: 137).

The group is the means to withstand the magic system. It serves as atmosphere. It brings elements for closing, filling of sense. The group appropriates the magical system. It sees in it something functional to correct dysfunctions of everyday life. It was considered prudent to speak of group, no community. Every community has groups. Some believe in magic, others are not, but are equally affected for it.

In short, magic as a system, it includes three types of information:

1. Information that is permanent or relatively stable for the performance of rituals.

2. Other information constantly renewed under certain laws of variation, consisting of the semiotics of the request, or if you prefer, dysfunctions: such as diseases, witchcrafts, evil salt, evil herbal and others.

3. The third information comes from the environment. It is represented by stories about effective magical actions occurred. They form the expertise. They give status and reliability to the magician.

We talk about renewal under the laws of variation because the semiotics of request is culturally conditioned. It retains certain constants perfectly determinable from a variation to another. In the magical request is not spoken about gastritis, for instance, but of witchcraft. The hex corresponds to a language and semiotics, previously articulated in advance into culture. The user only repeats few signs shaped by culture as a hex.

"Semiotics of certain diseases realizes of demons particularly complex and insidious. At certain times these demons-diseases act with ruthless and pandemic energy; remember the famous plague of leprosy, epilepsy and syphilis put a threshold to the Modern Age. These demons-diseases indicate clearly the corporeality of demons" (Giordano Bruno, 1982: 217).

The disease becomes a hex when it enters into the cultural sphere of magic. The user does not need to talk about organic symptoms. He only needs to refer to the features of the hex, which have been previously set by culture. The user reproduces the mental structure given to the hex. This is enough to communicate and be understood socially. With this, he can request the rectification of his place into society. He can claim an alignment of his body in order to become functional in his group. With the semiotics of hex, he can make a social criticism toward people causing the functional imbalance.

Classification of Magic

The strategy followed by several scholars to understand the complex phenomenon of magic, has been to subject it to classification. It has been considered as part of the occult sciences. In its classification they have taken different points of view. Some authors have given priority to the ethnic factor, others to historical development.

To Castiglioni (1947), the magic has gone through four major stages:

1. Prehistoric

2. Primitive

3. Natural

4. The divining

In prehistoric era, it sees magic as a natural response to the phenomena misunderstood by man. The primitive era covers the period of classic Greek, Indian and Egyptian culture. In this period would enter the magical manifestations of western and eastern culture, and indigenous cultures, including those still existing.

The natural magic took special interest in the West for its validation from the dominant neo-Platonic philosophy in the design of official knowledge of the epoch. This period rescued the ancient belief. The words and names were the reflection of the creative power of the ways from the mind of God. The old controversy from "Cratylus" by Plato was reconsidered. It was reconsidered the question about whether the name was on the thing or if it was independent of it. Large sections of the clergy and groups dedicated to the occult sciences chose to hold: the name was on the thing. With things like this, it was possible to act magically on the world and its beings. If the original name given by God to things was known, anyone could create the same thing. The task of the magician was to seek the original word and hide it of the profane man. Grimoires, full of clavicles, decipherable only to the initiates, were written.

In fact, the knowing of the sixteenth century does not suffer by lack of structure. On the contrary, we have seen how meticulous they are the settings of its space. This rigor is imposed by the relationship between magic and scholarship, not as accepted content, but as required forms. "The world is covered with signs that need to decrypt and these signs, which reveal similarities and affinities, are only forms of similarity. So, to know is to interpret: moving from the visible mark to what is said through it and that, without it, would remain like mute word, numbed between things" (Foucault, 1984: 40).

In this period, several renowned thinkers conceptualized about magic from philosophy:

Paracelsus, Agrippa, Pico Della Mirandola and Giordano Bruno, among the highlights. The latter was sentenced to the stake for his theory about magic. He classified it as:

1. Natural

2. Phantasmal

3. Extra-natural

4. Mathematics

5. Theurgy

6. Necromancy

7. Maleficent

8. Divinatory

These modalities grouped them into three broad categories:

1. Divine Magic

2. Physical Magic

3. Mathematical Magic

These three types correspond to the types of world identified by the author (1982: 225-230):

1. The archetypical

2. Physical

3. Rational

The magic of this period took the name natural. It tries to prove the existence of a different cause to from physical phenomena to cause an effect.

Natural magic represents another level of evolution of primitive magic toward experimental science. It leaves the magical fundamental concept as well as the religious idea, by using the rational form of philosophy, but remains linked to reality in its observations. It develops slowly from primitive magic. Observation comes and the method proceeds according to the force with which the critical faculties predominate over the emotional. It is attacked in the human mind the notion of supernatural beings, their intervention. It seeks to explain nature through its own phenomena. Natural magic became science, but retained for a long time, much of its ancestral heritage from primitive magic. It got hardly rid of some of its conceptions, especially those most deeply rooted in the deep strata of the mind.

The fourth stage of magic defined by Castiglioni, the divining, strives to systematize the magical process of divination. This period in terms of Eliphas Levi (1922), begins with the appearance of the Bohemians (Gypsies) in Europe. It is characterized by the unification of the methods of interpreting the tarot symbols. The effort was aimed at providing fixed content to symbols to eliminate contradictions and misrepresentations in reading and transmission of messages or mantic divination. Such an organization of the occult led to displace the trance of the magician as a central point of magical practices. From the most important thing so far, the trance became an element of equal status to others. The Magic acquired a popularizing never seen. It is no longer required special skills to exercise it. It was enough to memorize some fixed meanings for symbols.

In order to achieve greater clarity in the study of magical discourse, it should complement the classification by Castiglioni with an additional type of magic, this is the one experimental. Its beginning can be dated to the late nineteenth century, from the invention of the electronic media. These were important instruments for laboratory study of phenomena such as clairvoyance, hypnotism, extrasensory perceptions, the bio-field, and other phenomena considered as wonderful or unspeakable. In this perspective began to speak about parapsychology. This discipline was concerned with studying phenomena such as the paranormal (faculties Psy) not falling within psychology. It studies also the intellectual and mental processes without specific base, or from obscure origin, such as telepathy, clairvoyance, hypnotism, spiritualism and others.

It is surprising, these investigations began as marginal. Today received a great attention from the countries technologically developed. Starting assumptions are different from those of physics and the exact sciences.

To Rysl (1991: 78), research in the field of physics and other natural sciences has provided a uniform image of valid systems in our three-dimensional 'material world'. This image still has some gaps. Together, they have internal consistency. Explain, on a uniform basis, both: processes in remote stars as those that occurred on Earth. It explains chemical reactions of inanimate matter or biological laws of living organisms. The study of ESP (Extra-Sensory Perception) and PK (Psycho-kinesis), has demonstrated the existence in the world of systems not related to the laws of physics. The PK studies any influence exercised at a distance, not based on known physical forces.

Another classification of magic, perhaps the closest to a study of magical discourse by his semiotic features, is offered by Boriesson (1962: 14). This author prefers to adopt a classification based on the three essential forms of magic, according to the means used.

1. The mimetic or imitation magic.

2. The magic of enchantment.

3. The magic amulets, pentacles and talismans.

This classification of magic is based on the symbolic. In the first, they dominate the laws of symbolic thought expressed in the analogy and similarity. In the enchantment, it is predominated the fixed language of the magical texts. They reach for their fixity, the dimension of the symbolic. And in talismans, signs are elevated to the status of symbols. This is so because they are written in a material support to equip them with power.

It is also interesting to mention the classification of magic made by Uribe Escobar (1967: 39-40). In his text, he makes a collection of prayers and incantations from the oral tradition of Colombia, especially about the Department of Antioquia. It is not difficult to assume his influence on the formation of many local practitioners, given the popularity of his work.

The magic was the mother of all sciences in antiquity. Astrology and Alchemy were the basis of Hermetic Philosophy. At the same time, it was the progenitor of astronomy, physics, and chemistry. At the time of his mastery, magic was divided into the following areas:

1. White Magic or Theurgist

2. Black Magic or Arts Goetia

3. Natural Magic

4. Magic Talismanic

5. Cabbalistic Magic

6. Magic Theoretical

7. Practical Magic.

It is spoken about Natural Magic when it comes to the production of surprising phenomena and in appearance prodigious, by using only actions and means natural. These phenomena, because of their rarity, are out of reach of most people. The Ceremonial Magic considers ceremonies and operations relevant to works of invocation, evocation, spells, etc. The Talismanic Magic considers the making of talismans, amulets and other preparations of the same analogous species. The Magic Cabbalistic considers the general knowledge about Cabbala. It studies its operations and its practical procedures. The Theoretical Magic considers the doctrinal and philosophical content. And The Practical Magic is aimed to the experimental and scientific content.

Luis Carlos Molina Acevedo

Magic and Language

By studying the magical discourse, the concern can not be focused about if it is false or true. We are interested above all know what their inner laws are and what gives it existence in a community. The magical discourse can not be seen as an occurrence of a charlatan. It is too systematic to be charlatanism. At all times, it reflects the worldview of a social group.

If what was said by the magician, it is interpreted without regarding the voice of the community, present in every word spoken, then, we have an interpretation with a partial meaning. The language does not stay in the simple descriptions of the facts in which incur by part of some positivists positions. It should move towards the argument by the illocutionary acts. This allows the creation of properties for things. Without them, they could not have them.

Thus, we manufacture, for words, a new meaning, the type of representation. By doing this, we make really not 'things', but properties. They seem to actually belong to things. But, they are actually only justification from our discourses. "The illocutionary force shows us how to integrate or incorporate to the things those argumentative discourses that we do about them" (Ducrot, 1990: 48).

The magical action and his discourse, as such, can not be measured with criteria of truth or falsity. These are applicable only to the picture of the world represented by them. The study of magical action must rescue its self-reference. The polyphony of voices crosses it. Its social interaction is determined by the language. For it, its meaning is woven.

The magical discourse is linguistic expression of a social action. Upon entering into the magical system, it becomes an information packet. It initiates a process of reorganization of the altered information in both, the magician and the user. The magician can not remain untouched by the information received upon activation of the system. This information modifies or confirms its worldview. He is reaffirmed socially.

Magic and Symbolism

The language moves only in the field of verbal signs. The magic, besides these, involves nonverbal signs. Its study should also address the symbolic aspect of its practice. "Man is a symbolic animal" was the premise by Cassirer (1987: 48). He opened a wide field of research in the cultural as a sphere of knowledge, based on the symbolic. "The culture derives its specific character, and its intellectual value and moral, not from material that composes it, but from its form, its architectural structure" (1987: 83).

In the study of magical discourse, a special attention to the semantic classification of the symbols must be given. "The symbolic function is presented as a double movement in the subject: man does of his action an object, but to return to it at the right time its founder place. In this equivocal, operating at every moment, is based the whole progress of a function in which action and knowledge are alternated" (Lacan, 1981: 104). The symbolic function based the magical action. The possibility of cognitive action stems from the reorganization of information. This modifies the worldview of each participant in it.

The magic action is a work of symbolic score on the worldview of the user. With it is given meaning to him, or is renewed into him. It can also alter as in the case of witchcrafts. "It is a fact that is checked loosely in the practice of the texts of the symbolic writings, either the Bible or the Chinese canons: the lack of punctuation is in them a source of ambiguity, the score, once placed, sets the direction, its change renews it or upsets it and, if it is misleadingly, it is equivalent to alter it" (Lacan 1981: 131).

The magic has something of the structure of the poetic language. It operates on the hypothetical to bring in user a reorganization of the inside information in his state of soul, through management of the semiotic structure based on symbolic language. His sense refers to another meaning. Thus, it achieves the correcting of dysfunctions or fissures, of the cracks caused by the dissatisfaction of desire. The value of the magical text is not in his sense, but in his force, his potentiality, by outside of language. The magic text is no-language made language. It is a symbol made of language. It points to efficiency, to a vibration in the rhythm and musicality. So, it has poetic features.

The hierophanies, or symbolism of the sacred, addressed by the comparative history of religions, talk also about magic. In magic, it is acted with instruments. These are really symbols, such as the sword, magic wand, the crucible and the circle, among others. To operate them, they are separated from reality by the action of language and transported to the world of magic. "The world around us, civilized by the hand of man, does not acquire more validity than the one must alien prototype that served as a model. Man builds in accordance to an archetype" (Eliade, 1985: 63).

Luis Carlos Molina Acevedo

Field Research

Thus, symbols and texts of magic will give an explanation about the magic from the point of view linguistic and semiotic. It is a way of bringing together the two as representative elements of the American continent to explain each by the other. The fieldwork for this analysis was conducted in municipalities of south-western of Antioquia and Medellin, Colombia. In these places magic has four ways of expression, as follows:

1. Curerism (**heal-magical-action** or action of healing, quackery)

2. Herbalism (**herbal-magical-action** or action of healing with herbals)

3. Helperism (**help-magical-action** or action of healing with assisting of spirits)

4. Orisonerism (**orison-magical-action** or action of healing with orisons)

Each of these practices corresponds to an agent. They are named thus:

1. Curers (**healer-magician** or healers or quacks, men who heal with poultices)

2. Orisoners (**prayer-magician** or prayers or supplicants, men who heal with orisons)

3. Herbalers (**herbalist-magician** or herbalists, men who heal with herbals)

4. Helpers (**helper-magician** or assisted-men, men who heal with assisting of spirits)

These four manifestations of local magic have constant and recurrences with universal manifestations. His practice is no stranger to the basic features of the practiced in other countries in Latin America and the world. About these is this analysis. This will provide elements for understanding the magical phenomenon, regardless of its geographical manifestation.

Practitioners of these four expressions of magic drew from universal sources of magic, through grimoires and books of compilation. Often, this knowledge is not assimilated directly by reading. Many practitioners are illiterate. They have known the universal features of magic, through the teachings from experienced magicians. They saved for much of his lives to pay a course about magic. These were quite expensive.

PART TWO:
CHARACTERISTICS OF MAGIC

Luis Carlos Molina Acevedo

The Magic

The study of magic must be addressed in a three-dimensional sense, as follows:

1. Human Action

2. Historical Event

3. Social Phenomenon

Thus, it corresponds to every object of human sciences. In turn, the magic is possible to identify three types of discourse:

1. Fixed texts generally called 'magic formula'

2. Narrative texts to ensure the permanence of magic

3. Meta-texts to provide a philosophical basis to magic

The social movement of the three types of magical discourse is variable. The narrative texts are the most social transaction. These are follower by the magic formulas. And in a more restricted circle, they are passed the meta-texts.

The magic endows man with will, in addition to repair and maintain its own forces worn by use. "Man, that influences the nature through action over the human beings, thanks to the word, and that rises to God through prayer and ecstasy, is the bond that unites the Creator with all creation" (Papus, 1989: 107). In the magical model, the world is an object. The magician can manipulate it at will.

The man is manifested in the Universe by the action of the will. It allows him to fight fate. He makes of this an instrument of his conceptions. In the act of imposing the decrees of his will to the outside world, man is perfectly free to go to the lights of providence. Or to ignore the one designated by it. "No fact, however simple, it is no longer the translation that the nature makes about a principle that is emanated from God, and man always can restore the link that connects the visible fact with the invisible principle, what contains the enunciation of a law" (Papus, 1989: 108).

Qualitative Description

"In this town, there are a lot who do the damage, but few that remedy it"; it is the phrase of everyday language in the municipality of Fredonia and in many other municipalities. It seems to sum up the magical phenomenon in many municipalities of Antioquia, Colombia. It reveals an accepted reality. Each of its inhabitants is a potential practitioner for personal purposes. Few are engaged in a social function of it. Most patients for cases not related with prayer-magician, should go to neighbouring towns or to the capital of the department to seek a magician. This lack of practitioners is only apparent. Practitioners of the town do not serve people the same. The clientele of them comes from outside. They must be recommended by acquaintances or patients already treated as requirement to serve them. With this, it is absconded the moral sanction and social. Nobody wants to be associated with witchcraft practices.

Most practitioners learned their art by oral tradition. They have read books on magic, but it is the knowledge transmitted the real trainer. They sought teachings of practitioners recognized in Colombian departments as Chocó, Valle, Putumayo. They did also in municipalities of Antioquia as Remedios, Zaragoza and Segovia. They brighten: the essential goal of travelling to these places, it is not knowledge. The important thing was to make contact with experienced magicians. They passed them power. They made the rite of initiation. They were invested with magic power. This last point is crucial. It reflects a fundamental assumption. People without special gift for magic, should receive a transmission of power from a great practitioner. Without it, social activities would lack effectiveness.

Cazeneuve (1971: 164) identifies four procedures for black magic:

1. Proteiforms of Witches.

2. The transfiguration or lycanthropy of mohanes in pigs or bunches of bananas.

3. The split in the case of the goblins.

4. The association or pacts with supernatural beings in the case of helper-magician.

These forms are not uncommon in places subject of this analysis. It is surprised how are repeated the magical ways in different places, outside the municipalities object of this study. It leaves the feeling to sit reading a national author, rather than a foreign one. These coincidences denote the universal constants present in magic.

As for the transmission of power, it may be from parents to children, including relatives, indigenous to ordinary person, or the magician to a trusted person. When a person has a natural gift for magic, usually is an experienced and socially recognized magician who finds out him. He needs be recognized socially by someone. Depending on the type of procedures to achieve the magical action, the initiate becomes:

1. Prayer-magician: acts with orisons.

2. Healer-magician: acts with poultices.

3. Herbalist-magician: acts with herbs.

4. Helper-magician: acts with the help of supernatural beings.

Luis Carlos Molina Acevedo

Magical Activities

The **prayer-magician** directs his magic action to relieve sprains, broken bones, bleeding and other physical ailments in humans or animals. His place of serving is public places such as bars, cafes and ice cream parlors where people are looking for them. This is indicative of social acceptance. The other practices not reach this acceptance. Those have activities of work different to their art. The prayer-magician keep otherwise of earning daily sustenance. No charge consultations. In some cases they left to the will of the people any compensation. The ritual is marked by passes of hands. They are drawn crosses on the ailment while orisons are recited. Of these, the user only captures a slight murmur without decrypting the content. The sense of magic act derives of ritual as a whole. People see in the orisons the possibility to relieve their ailments without suffering pain at the same time to be cured.

The rituals of the **healer-magician** vary from practitioner to practitioner. The one made by a healer-woman from the municipality of Fredonia, gives sufficient elements to observe their characteristics. She asks the user to be accompanied by relatives. She asked several chicken eggs. The appointment begins with a diagnosis where glass vacuum valves are used. They contain liquid of colours. The patient holds one with his hands. If the liquid expands and bubbles, he is ill. In this case, he is led to a room. He is lying on a bed. Some relatives are located. Incense is burned. The healer-woman extends ribbons of colours on the patient's body. Then, she spikes needles of different sizes in the patient's body. All these acts are accompanied by the recitation of orisons. The people, present in the place, only capture the murmur of her. After the preparation, the patient drinks an emetic. Then, the healer-woman opens small holes at the poles of the eggs with a pin. With them ready, she sucks with his mouth in the mouth of the user through eggs to force him to shed the curse. The eggs are used like straws. She asks the patient that he such content ejects it blowing through eggs. The eggs sucked and blown are deposited in a vessel. The operation is repeated until when the healer-woman considered that the evil was expelled. At this point, aromatic drinks are given to the sick. He is cleaned to go to search the vessel with the egg bakes. One by one is examined to find the jinx. It may consist of a small tube with a plume inside, or coiled hair and larger than usual. Then, the healer-woman explains how the evil was put into the user. She says, among other things, for this, they took a membrane or intestinal villi removed from the cattle. In this, it was introduced the hair or plume. Then, they gave it in some food. This capsule is joined at the stomach wall to perforate it, feeding on the tissues. Hair or plume grew like any living being to produce the ulcer cause of evil. In this sense, the suction was necessary to avoid the return of the capsule. If do not, it is implanted elsewhere. Then, it causes more harm. It should be clarified that the use of needles and the mixture of the ribbons, according to the colours, varies according to disease. The semantics of the healer-action comes also from the ritual as a whole.

In the **herbalist-magician**, the efficacy of ritual focuses on the collection and combination of elements. With these, drugs or potions are obtained. The most important semantics here comes from atmospheric conditions. The position of the stars is considered. They set the time, day and month of year when certain specific plants should be collected. The same conditions determine their mix. For example, some people point out the importance of having roses at home to reach the loving harmony. To cut these, the person should be prayed an orison to elemental or spirit of the rose. Thus, it does not suffer and accept for fostering such a state of loving harmony. Here, the ritual is not applied to the user, but on the elements for treatment.

The rituals of **helper-magician** are presented as the most complex and symbolic. Its purpose is to invoke the supernatural beings. So, they agree to help the magician. As an example, one can cite the one followed to maximize the monicongo (golem-puppet). After the ritual, the object concentrates supernatural powers. They are put in the service of helper-magician. Through them, perform his magic actions. The golem-puppet is a doll about three or four centimetres tall. It is manufactured with virgin wax removed from the cells not used in bumblebee hives.

This insect is similar to the bee, but larger. Such figurine buried for three days with a special ritual. It includes to prayer specific orisons. Then, it will be unearthed with another ritual. Finally, it will be saved in the carriel (big leather wallet slung over the shoulder). Since then, the golem-puppet is obliged to do everything that is requested by the magician. This, in turn, will feed it with needles every day. In this regard, it is important to note some convergences in the use of similar objects through history and in different cultures. Even, it comes together in the idea of the Golem of Jewish culture, or the Adam made with red clay by alchemist.

Magical Figurines

"Ooh thou, magical figurine, listen to me! I have been summoned, I have been sentenced to perform work of all kinds, those which force doing the spirits of the dead in the afterlife; you know then, ooh magical figurine: since you now possess longer useful, you must obey the man in his need! You know therefore that you will be the convicted in my place, by Duat watchers: to sow the fields, canals filled with water, carry sand from east to west ... (Statuette answers :) —Here, you have me ... I await your orders ..." (Bergua, 1962: 77). Perhaps the concept of the golem-puppets comes from this passage from the "Book of the Dead". It describes in perfectly what is the purpose of these statuettes.

Among the Chinese talismans, it should be noted figurines of clay, paper, wood or jade, manufactured by the witches Tao-Niu. To insufflate life, the witch does not conform to enclose inside reproductions of vital organs: heart, liver, lungs. It introduces also a live animal, bird, insect, and reptile. The spirit emerges from the animal. Since the time of his death, it will dwelleth in the talisman.

"It is a rite of animation of a statue, which find also different details, in the rite of resuscitation own of the Egyptian magic funeral, in the Hebrew Golem, in the Thayphap of chaff from magicians Anmitas" (Boriesson, 1962 : 88).

Across Asia, they are encountered the same belief. We find it in Egypt. In Israel, the Golem crystallizes the magical power of statues. His creator gives them life breaths. Among the Hindus, this rite is called prana-pratishtha. It is intended to convey, through psychic emanations, the energy from adorer to the inanimate object. The life, breathed in this latter, is maintained by the daily worship.

To Barylko (1977: 33), the old dream of Cabbalists was to know the hidden name of God to work wonders with it. And of all the wonders, the most would make a clay body. It sought join it the God's name to breathe life. Finally, to create a man, and thus, the magician equated with the same creator.

These convergences throughout the history of magic are sufficient to show the great aspiration of magician. One of the great dreams of a magician has been the one to create a replica of the man to put it at his service. This would free him from his destiny to earn a living with work. These are not the only semantic convergences. Within this isomorphism, it is also entered the homunculus or androgynous of the alchemists, Zombies of the ritual voodoo in Haiti, the Android of Christian cabbala, and even the current automaton or robot of cybernetics. That is, the same semantic content expresses a common aspiration to all cultures, but is objectified differently. This aspiration of magic has also become the aspiration of technology. Technology seems the modern name to magic. It seeks a mechanical man capable of acting as a human being. It is named Android, the same named given by Christian cabbala to its golem-puppet.

The writer of Antioquia, Tomás Carrasquilla, in his book "La Marquesa de Yolombó" (The marquise of Yolombo), speaks of dependents or golem-puppets. "The dependent is a doll, something small, very funny and very flattering, that one leads and not lets that to one happens him bad thing and that ends very well all that one does and in all what one undertakes" (1974 : 253). And about his preparation, he says:

"One sends him to do to who knows lifting and pays it! One bought is not good, because they put one pig in a poke. It has to tilling a magician that knows the 21 hidden words, which have the power to help and liberate. He says them in the ear, the dependent, when he has it finished and painted. And it has already virtue. It is tilled from the root of a stick, which is only known by the magician; and painted with a black dye, which no one knows, but the magician does know. They have to take the root from deep into the earth and during night, and at night should be tilled, because a dependent is not serving, if it sees the sunlight, before or after of being made. Who leads the Dependent has to see it with a candle, even during day; and cover it and wrap it, if one has to put it to a side, to change his clothes or to bathe, and it can not be shown to anyone, either day or night" (1974: 255).

Magical Lexis

To better understand the magical phenomena, we must know its semantic and lexical fields. Then, we talk about some magical terms in the oral tradition.

Healer-magician, this term was initially appointed who would cure bites of snake and poisonous animals. Then, it took a broad designation, as the act of healing covered to people and objects. An object can be cured to provide protection to its owner. It can also cure the body itself, i.e., be sent to close the body, so you do not enter any wrong in life. Of course, at the time of death, one will suffer and rest only if someone opens a wound on the sole of his left foot. Cured objects are the named **contras** (protections against snakes), or if they are jewels, these are named **amulets**, and if they carry inscription of magical characters are named **talismans**.

The healer-magician cares of having to the hand the contras. Usually, they are black, with mineral consistency. They are stones supposedly obtained from snakes, or removed from the black cat when is done the ritual described in the **grimoire**, book of magic formulas. He manages also "Chupaderas" (**poultice**) to absorb the poison from snake bites. They are made from cane sugar, snuff, or may be purchased from the indigenes. The large arsenal of knowledge of healer-magician comes from the **secret-substances**. This is the name given to inscription or essential text of this kind of magical discourse.

The **prayer-magician** depends for his magical activity, knowledge about certain **secret-orisons**, which is the name given to the inscription of this type of magical discourse.

The **herbalist-magician** bases his magical work in the power of plants. He learns to manipulate it through **secret-recipes**. These are inscriptions of language to operate with the plants. With these, they are prepared **concoctions**. They are the generic name for a series of natural preparations for multiple purposes. With these, they are also prepared **love-philtres**. In addition to plant components, these philtres have also animal ingredients and elements or secretions from the person that one wants to love. They can also do **herbal-enchanting**. Contrary to the above, it seeks to stupefy the spouse with herbal-recipes to have lovers without problems. This person is named **enchanted-man**. But just as there are those who put them there are also who may cures them. This is the main social function of the herbalist-magician. The **potions** are vegetable compositions for minor ailments. **Herbal-baths** are obtained from the decoction of various plants. With them, the person baths to have health, money and love. His mixture is given as seven cabalistic numbers. The **herbal-irrigations** are similar to the herbal-baths, but this is used to recuperate the business went down the drain, remove bad luck. These are used to remove the **evil-salt**.

The **helper-magician** uses supernatural powers to perform his magical work. They can be **hexes**, consisting of ailments. With these, people go consuming, even until causing death. Hexes can be objectified in **snakes**, **frogs**, **lizards** and similar animals. They make their nest in the stomach of the victim. They can also be nested **worms** anywhere in the body. The **evil-eye**, whose existence seems to be universal, is a durable effect in children, caused by look of a black magician. It is manifested with water jugs (rash) throughout the body. The helper-magician can also put the carate or white patches, vitiligo. They appear on the skin by using frog bones to do evil. The power of helper-magician is focused on the **golem-puppets**. These puppets harvest coffee, fight and triumph for him at all.

Among the helper-persons, they are also placed the **witches**. These are **enchantress-witches** when engaged in making **love potions** using menstruation, semen, hair, nails and others components. They are **flightless-witches** when engaged in flying the haunches of the devil in the form of buzzard or fly. The helper-magician can also be a **goblin** to chase girls at night. He lets her sucked and scratches as evidence of his visit. Sometimes, he inscribed characters with his nails in the body of the victim. This goblin is different from **goblin-rapper**. It makes mischief at night in homes. They drop things to the floor to wake the sleepers. When these stand, things are all in place. The **goblin-child** is a fallen angel. He likes mislead children. The helper-magician can also become a **Mohan** to avoid difficult situations. Thus, during the time of smuggling liquor and snuff, it became bunch of bananas to mislead guard's income. They can also be converted into pigs when are without money. They ask someone that sold them without the lace. When the sale is finalized, they disappear of sight of the buyer. This keeps with the lace in hand. Mohan appears shirtless in the place of sale.

The helper-magician in his works of **black-magic**, use the **magic-circle** to protect him. This occurs especially when he seeks the **fern-flower**, on the eve from the feast of San Juan. This will grant greater powers to its possessor. The helper-magician is who uses the photograph of a person. It forces her to fall in love by who demand his services. He makes also **vigils** when he wants to do evil to another. For this, he places the photograph of that person like a coffin and around, lit candles. Secret prayers he prays. He **join-up** people to love each other forever. For this, he makes a symbolic ritual of joining objects from them. These must have been in contact with both people. He **binds** one person to another by the operation of casting and prays the orison about seven knots. He makes them both on a string, hair or the like. The helper-magician **smokes snuff**. With this magical operation, he returns the absent one to where the loved being. He will use the **clairvoyance of glass** of water for seeing at distance. The user can see, in this, to who did him the evil. The culprit is reflected in the glass of water.

The helper-magician will go to the **land of cemetery** to evil-salt homes or businesses. He does this when people do not like him or they do not sympathize with him. He uses also the **bones of dead** for the same purpose. Here the ritual is different. The same bones are used to look after the house when for some reason should be left alone. The attackers always see the house full of people. Really see the **souls of the dead** whom have those bones in live. With bones of dead are also made dice to always win in the game.

"The main purpose pursued in this operation when practiced with the skulls of dead bodies, as evidenced by the numerous perforated skulls found in Europe, Africa and Central America, was that of carving bone shaped disc that could be used as amulets. The possession of bones, and mainly the skull of one of the forerunners, provides a special force because provides the holder with the full force and power that belonged to the dead" (Castiglioni, 1947: 53).

The aided by the effect of **bi-location** he can travel in space. He can be in several places at once. By **clairvoyance** he can travel back in time. He can know the past or future things by the power of concentration. As for the objects of language, the helper-magician has **conjurations** to command the higher spirits performing the desired action. With the **sortilege**, he can invoke the intervention of an evil spirit. With **prayers**, he may request the intervention of a good spirit or the saints. With **incantations**, he can alleviate bad. With **exorcisms**, he can eject a possessed spirit of a person.

Magic, in turn, can be **white** when it comes to religious media. It is **black**, when the demon is invoked to do evil. A mixture of the both produces the **grey** magic. The **red** magic means all matters of love. And the **green** magic relates with the knowledge of plants and their virtues.

To **heaven** go people who were good on earth to enjoy the presence of God, angels and saints. **Hell** is presented as opposite to heaven: black, with permanent fire, place of suffering, roads full of thorns and inhabited by the animals to which one was afraid when one was alive. And **purgatory** is assimilated to a prison where a penalty is paid before going to heaven.

The saints are the intermediaries between man and God. Some priests are seen endowed with powers to heal. Many people intuit the miracle has the power of God. Magic, instead, with the help of the devil. **Holy water** looks like a great element to ward off demons of houses and the bed of the dying. It is also used to get out evil-salt from business and dwelling places, take away bad luck. With **medals of San Benito**, they are take away bad neighbours. The **ends of stolen candle** have the power to help carry out domestic work magic. The **Agnus Dei** (piece of candle blessed by the Pope, according to popular belief), is loaded as one of the best talismans to ward off the danger. The **devil** has the image of the ugly black with tail and horns. Envy is the leading cause of evil. The hope of good things always emerges in the phrase "faith will save you."

The words in bold in this chapter constitute the lexical field of magic.

Magical Symbolism

Gilbert Durand (1981) makes a classification of symbols. The symbols are grouped into constellations. These are grouped around three main archetypes. He represents them with the symbols of Tarot:

The first with the scepter and the spades

The second with the cup and its contents

The third with the denarius (diamonds) and germinated rod (coarse or clubs)

Similarly, in the magical ritual can be identified symbols. They are located within these three kinds of archetypes, according to its semantics.

Magic has as the main ceremonial symbol to the **magical circle** (see Figure 1 at the end of this book). Within this, the magician is placed before the invocations. It is drawn with charcoal. "Every magical operation must be performed within the perimeter of a circle that symbolizes the will of the operator and defends him from all outside bad influence" (Papus, 1988: 413). He describes his path. The circle should be nine feet in diameter where there are three concentric circles separated by a span. In the centre circle shall be the name of the time of the operation, the name of the angel of the time, the label of angel of the hour, angel names and ministers. The magic circle is a symbol full of symbolic language. It is important to the operation on time, entailed either by the circle. This must be synchronized for the ritual like clockwork. For this correlation tables are used, which group the symbols of time, as will be seen later. One could say, essentially magic operates over time. It is a time machine to stay, back or forward depending on the need for the task.

Illustration, taken from Tondriau (1966: 324), represents the circle of invocations of Eliphas Levi. For a better understanding of his explanation, the reader is referred to Figure 1, at the end of this book.

Another important symbol for the magical operation is the **scepter** or **magic wand** (see Figure 2 at the end of this book). With it, the magician can project his power over the world to dominate it. By the magic wand, the magician power in the world is installed. This symbol, next to the sword, will be the first archetype of the classification of Durand.

"To indicate and to direct the projection of the will, the magus has at his disposal an instrument built with wood and called magnetic iron rod or magnetic scepter. This tool of magus has no other object than to condense a large amount of fluid emanating from the operator, or the substances that prepares for the purpose, and direct projection of the fluid on a certain point" (Papus, 1988: 172).

With the sword, the magician defends from spectres and spirits enemies. They presented at the time of invocation to disrupt the ritual. The sword, by semantics of the stabbing, can dissolve the larvae. The black magicians cast them from his astral body.

The sword "is intended to serve as a defence for the operator, and the tip ends owes all its qualities ... The fluid conglomerations formed by the union of an astral power acting as the soul, the vital fluids of the environment acting as body, have a great analogy with electrical conglomerates. The astral can not influence the physical, but through fluids of physical life, we could say the vital electricity. So, when the operator assumes that the astral power that he has, he wants to abuse his power to oppose the intended purpose, the operator has no other recourse but to present the point of a gun to the fluidic entity presented. The metal tip instantly draws the astral-electric effluvia that constitute the body of the animated being of evil intentions and suddenly, this being feels stripped of all its means of action on the physical plane" (Papus, 1988: 172).

It is particularly significant the popular practice to catch witches. A needle with its eye nailed in wood of the door. It is left out the tip. The witch can not exit. Facts like these show the universal scope of magical practice. Beside these sharp instruments, they are also: the **white-handled knife**, **black-handled knife**, the **dagger**, the **lancet** and the **hook**, among others. Each carries a specific symbolism.

The second archetype is represented by several magic symbols. Here are the **containers**. In them the compounds are prepared. The **crucible** serves to melt metals. As the magician must be purified before rituals, the **water** comes to act as content. It is consecrated with the recitation of psalms of the Bible (26, 13, 38, 68, and 105). Thus, it acquires the semantic isomorphism of the continent. The **magician** to get into it, he becomes content with the recitation of psalms (50 and 23).

Beside these symbols in where symbol constellations are grouped into archetypes, it is possible a more detailed discrimination of symbols. In theriomorphic symbols, for example, they are: **toads**, **tadpoles**, **snakes**, **bats**, **worms**, **owls**, **dogs**, **cats** and many others, present in the magical arts. Papus clarifies well the functionality of the same in the ritual, and the use of them for specific jobs.

"Animals are used in magic as eliminators of astral fluid is needed to effect some operations ... Witchcraft is based entirely on the principle of animation of consecrated objects, animation that is obtained thanks to the astral body of toad which it is set in the aforementioned objects ... if need be replaced one of the disciples in evoking, it can be done by a dog, animal whose magnetic aura is very powerful ... the feathers of birds in accordance with plants, are used as swab to mirror the water energized by the magnetic influence" (Papus, 1988: 220).

But the symbolic power from these animals, there are not in themselves. It is in the operation to which they are subjected by the magician, according to certain planetary conditions. The following classification of correlations is quite illustrative:

1. **Planet**: Saturn, Jupiter, Mars, Sun, Venus, Mercury, Moon.

2. **Bird**: Hoopoe, eagle, raven, swan, dove, stork, owl.

3. **Quadruped**: bull (toad), deer, wolf, lion, goat, monkey, and cat.

4. **Fish**: cuttlefish, dolphin, licium, thimallus, sea cow, trails, Lurus.

Correlation tables are tables. Each term in bold, it leads a column. The elements in each column are ordered. This way you can replace a term by any of the other columns, if it is in the same position of order. For example, dolphin (in the second position in the fourth column) can be changed by deer, eagle or Jupiter (occupying the second position in the other columns). Here the correspondences are shown as lists to facilitate the design of the book, especially in digital versions, but the original correspondences are tables.

Generally speaking, the instruments used for magical operation, they interest primarily by its symbolism, besides the power foisted on them with the ritual of the magician. Yet, they are subject to grammatical rules which have a function of categories. Each of them can be replaced paradigmatically in the linear combination of operations, in accordance with the correlation tables. These paradigmatic substitutions in the magic circle, for example, are intended to operate over time, as noted above.

The above classification is only a sample of the many correspondences that exist in the magical discourse. The symbols are multiplied exponentially with each new box of matches. These tables are intended to help the magician to use the appropriate ritual objects. The magician can go to any of them when doing a job. For example, to represent the moon in a ritual, he can go to an owl, a cat or Lurus. If the ritual should be done under the planetary influence of the moon, the magician can use an owl or a cat, or Lurus, to represent the moon, and with it, the planetary energy required to bring it to fruition.

The importance, of correlation tables, is also given by the universal scope given to magic. Through these tables, the magician is guaranteed to have the appropriate elements for the ritual, independent of the variables of time and space. For example, there will be elements of spring, but the ritual is performed in summer. The correlation table ensures infallible replacement of a spring element by another of summer without affecting the efficacy of ritual. So, the temporary variables are managed.

In the case of spatial variables, correlation tables guarantee also universality. For example, an essential element for the ritual is produced in Europe. With correlation tables, it can be replaced by another produced in America, without altering the effectiveness of magic. In another field of human activity, hardly, it can be ensured such universality.

PART THREE: LEGITIMIZING DISCOURSE

Luis Carlos Molina Acevedo

Classification

Bibliographical sources of magic are classified into two:

1. Grimoires

2. Philosophical Treaties

The grimoires have, as main function, the one of validating and spreading the magic discourse. Philosophical treatises about magic have, as main function, the one of explaining the magical discourse. These validate also it because can only be explained the existing one.

Luis Carlos Molina Acevedo

Grimoires

"What we now think magic is usually the medieval European tradition whose most important relationships we have still more or less in original form. The Grand Grimoire, The Grimoire of Honorius, The Arbatel of Magic, The Lemegeton, The Great Albert, The Key of Solomon ... What majestically roll these titles ... Unfortunately, the content of these magical texts are mostly recipes to evoke the devil or one of his assistants with the majestic fiendish purpose of conducting the magician to buried treasure. Books with Solomon in the title, they are another powerful current of medieval magic ... The Cabbala" (Freedland, 1973: 148).

Apparently, these works opened the magical activity in the West. It was the time when Christianity was consolidated as a universal religion. It was given step to the monarchical form of government. This accounted for the progress of magic, going from an eclecticism of doctrines where the principles of different cultures mingled in front of the subject, toward an authentic Western magical setting. The monarch was the possessor of divine power, which received an inheritance. If someone else wanted to have power over the world, he should look at other sources than divine, exclusive of the king. The devil, then, became an effective alternative for commoners. The master of darkness became an object of worship, with the consequent production of grimorios where described worship to the devil and his servants willing to grant all kinds of favours.

It arouses curiosity the controversy brought from the history of magic around if Pope Honorius III and San Cyprian were the authors of the books "Grimoire of Pope Honorius" and the "Book of San Cyprian" respectively. Faced with the latter, Eliphas Levi does not rule out that St. Cyprian has left some manuscripts to his life of magician, previous to his conversion to Christianity. As for the first, he regards it unlikely. Pope Honorius III (Cencio Savelli), crowned pope in 1216, was one of the most zealous pontiffs of the thirteenth century. It was he who confirmed the Order of Santo Domingo, besides preaching the Crusades. In any case, the controversy points to show the concern of the time, about how the Church might be involved in such works.

Papus (1922: 287) says about the grimoires: "At first glance looks like a tissue of absurdities; but for those initiated into the signs and cabalistic secrets, is a true monument to human perversity; in it the devil is presented as a powerful tool."

To show the possible connection of the Church with magic, adds, "At present, it is impossible to find it (the Book of San Cyprian), said the merchant. The last one I had in my hands, I sold a priest for a hundred francs" (1975: 106).

Magic: Symbols and Texts of Magic

One should still throw a glance at the grimoires, marketed locally. It is not easy to establish how faithful are to the original in content. But they present their content in an indisputable source for magical discourse, especially when somehow preserve the tradition of magic in the West. This interest is not in fidelity to tradition, but on the amount of fixed texts and symbols included in these books. Somehow, they have shaped the worldview of the local magician. It has become a significant cultural form of our communities.

Among the titles that are available in cheap editions include: "The infernal book", "The Great Book of Saint Cyprian," "Enchiridiones, Grimoires and Pentacles", "The cigarette ash," "Secrets and Recipes", "The Holy Caravaca cross", "Mystic Crown", "Secrets of Nature" (most taken from the book of Great Alberto), "White Magic", "Black Magic" and the "Magician Chain", among others.

They should also be mentioned some titles of restricted acquisition as the "Red Dragon", "The Black Monk", "The Black Aura" and the "Book of Magic". One can intuit that cheap books are fragmentary compilations of the original versions. Instead, restricted marketing, hopefully, be more faithful to the original versions. Restricted versions are considered bibles true about magic where learned to people born from the second decade of the twentieth century. Only the initiates had access to them. This was done without distinction of social class. Proof of this is the amount of books on secrecy, found in the collections donated to the National Library of Colombia. These donations were made after the death of one of the important members of wealthy families in the country.

By that time, the common interest of training, especially for the lower classes, was to study two things in life. One was the "fencing" or the art of fighting with machetes. And the other was magic. They saved for much of his life to pay a course on the subject of interest, since they were worthy of imitating social models. Machete art earned him admiration and respect of others, who dominated it. The magic allowed him to be feared or envied for power to operate on the world and its creatures.

All these books have common themes. It seems as if they were copied from each other. Therefore, as a general observation it can be said, these works are neat and structured as if following the principle of everything in its place and every place in its thing. These works seem crystallization of the great dream of Giordano Bruno's art of memory as a change of mind, where there is a place for everything. "Already in Camillo, The Art of Memory was the place, where the theatre is exposed and gives the face. But the face with the mind is presented, and thus the universe, in the artifices of Bruno, is simultaneously image, monster and talisman; it is both artifice of the rhetoric word and artifice of the hermetic word" (Gomez de Liano, editor Bruno 1981: 292). One could see in the magic the same germ of post-modernity. At least magic as activity of imagination has always been related to the artificiality, to the monstrous, to the mysterious, in addition to its marked tendency towards the negative as opposed by to the positive into modernity.

"The art of mind is in Bruno not only an attempt to manufacture an artificial mind, and give it a theatre and a city that serves as a spectacle and transparent cabin itself, but is also an indication of the process aimed at the magic unification of understanding and to his confusion with the mentality. The whole arsenal of images or forms that populate the locatives joints of this mind are just tuning, as 'composition of place', of the spiritual exercises that would do possible the reform of the mind ... Imagination is also, as Synesius and Bruno said appointment, the vehicle of the soul and the body that mediates between the temporal and the eternal" (Gomez de Liano, editor Bruno 1981: 293).

It is as if the post-modernity had always been germinating in humans or at least, the one named by this word. It is as if it had always been part of the transit of man in this world. But now, it is expressed more intensely. Their features are used to define an era. This would explain why magic has been and will be it of all times and peoples. It is an expression of indeterminate need for immortality. In this, one can be something other than who is. "There is a certain magic that allows us to transmute and transform reality according to our needs indeterminate, purely psychic, i.e., those that have no practical satisfaction" (Molina, 1990: 42).

Anyway, beyond the surface of the contents present in the grimoires as magical discourse, one should read the great scenery, great performance, great theatre where the man lifts his mind through memory techniques as a hermetic art. So, in all grimorios appear clavicles, or keys, or pentacles of

Solomon where the keys are encrypted to defeat the Sphinx, the ineffable. The Sphinx as a symbol of time in continuous return is a fundamental symbol of magic (see Figure 4 at the end of this book). The Sphinx is the time machine for magic. It rules the vital watch of the creatures in this world. The clavicles are the keys to access the numinous, the beyond. But, they are also the crystallization of ancient knowledge, the magical knowledge, that is, the knowledge par excellence. The pentacles are those seals where all knowledge of man is stored, decipherable only through a process of initiation into the secret knowledge of symbols or correlation tables.

By seals understands Bruno (1981: 297) signs, diagrams or magical figures, used as places to store and restore the memory. They are not only based in the rules of traditional mnemonic technique. At the same time are signs of mathematical magic. They contain great occult powers. They elevate finite memory to the height of the divine mind.

Grimoires are objectified language, whose constituent signs are other objectified language in seals or pentacles. They are symbols of the archetype of the deniers (gold stick or diamonds). They evoke the wheel as a universal symbol of the cyclical, of becoming, the myth of the eternal return, and thus, of the immortality. Grimoires are objectified language from another previously objectified language.

Magic Treaties

The magical discourse turned a hermetic art. This led to the publication of philosophical works devoted to unravel the meanings of these objectified languages. This was made in the light of art of the cabbala, while superior sacerdotal art. It still retains the constituents of the procedures elided in the grimoires. These, in the hands of inexperienced, ended up being pure superstition.

The Cabbala is the Jewish tradition of the occult. It is a more philosophical attitude, not a technique to make magic. The Jews, expelled from their homeland for nearly two thousand years, had nothing but their holy book, the Old Testament, to which adhere. It was made comment after comment, as the Talmud and the Mishnah (first part of the Talmud). Under the circumstances, all the creative force of a people focused on the exposure of a book. Soon, they began to see hidden symbolic meanings behind the words. "To a large extent, the history of the grimoires is the one of 'ancient Hebrew magic', but falsified" (Freedland, 1973: 148).

Cabbala languages are just wrapping a powerful and sublime content. It goes from eternity and goes to eternity. Each Cabbalist used books, arrived in his hands, both Israelites and other peoples. Of them borrowed formulations to convey to the reader some mystery revealed. "If one used books from platonic origin and other books from Aristotelian origin, does not mean it that Cabbala, itself, is Platonic or Aristotelian, but each reached for a specific conceptual garb to communicate their personal intuition" (Hillel Zeitlin quoted by Barylko, 1977: 6).

Cabbala, despite having Jewish origin, soon had extensions to Christianity and Islam. "In the Renaissance, the cabbala is entered in various ways to the bosom of Christian meditation and, even, arises a stream known as Christian cabbala" (Barylko, 1977: 42). In this current are located philosophical works. They seek to unravel the hidden meanings of magical discourse in the West. Among its most prominent authors are Eliphas Levi and Papus. "The twentieth century's greatest magician, the Frenchman Alphonse Louis Constant, thought it was appropriate to take the Hebraic pseudonym Eliphas Levi for his work "Transcendental Magic", who unified for the first time the concepts of the old grimoires in modern terms" (Freedland, 1073 : 148).

Anyway it must be stated, Eliphas Levi is the nineteenth century. His production begins when he is expelled from the monastery where he received an influenced training by the frustrations of the French Revolution. It was the advent of social utopias as a refuge from the social ideals of popular type betrayed by the middle class. At the end of his "History of

Magic" he says: "Our history of magic has been intended to show that, in principle, the great symbols of religion have been at the same time the science then hidden" (1922, 487). Eliphas Levi's effort is directed to rescue the privileged place of magic and science of interpretation. His academic preparation is directed to free it of superstitions.

Men and things are magnetized by light like the suns. By electromagnetic chains, whose tension it is caused by sympathies and affinities, are able to communicate with each other end to end of the world, stroke or hit, hurt or heal, in form natural, but invisible, of prodigious nature.

That is the secret of magic.

Magic, it is the science that came to us from the Magicians.

Magic, it is the first of the sciences.

Magic, it is the holiest science, because it sets in the highest form, the highest religious truths.

Magic, it is the most maligned of all, because the world stubbornly confused magic with superstitious witchcraft, whose abominable practices we have denounced.

"It is only by the magic that can be answered the enigmatic questions of the Sphinx of Thebes and find the answer to these problems of religious history, sealed in darkness and sometimes scandalous found in the stories of the Bible" (1979: 119).

In the philosophical conception of magic, it must be regarded as practical science in the occult sciences. In contrast to the positive sciences, this proceeds by synthesis, not analysis or segmentation. In this perspective, the myth is considered as the expression of a hermetic knowledge. Initiates hid it of the common people through the coded language of allegory. It can be decrypted by the knowledge of the cabbala and its correlation tables.

In the interpretation of the allegories, the largest contribution comes from Papus (Dr. G. Encausse), who from his medical condition made a great effort to systematize the methods of interpretation in magic. For now, it interests as a source of magical discourse, in especial, his masterly exposition on the expression of ideas. He explains how to access to them by the initiate to interpret and decipher the objectified language of books, pentacles, talismans and magic formulas. To do this, he takes the phrase "The child needs a father and a mother." With it shows the degree of complexity assumed by magic to express any idea. So, he comes to the one here so-called magic: **objectified language by the imagination to work with power over the world**.

Papus, in his study about the language used by the magic, reveals how encryption of ideas is achieved. It is quite didactic about it. For him, the initiate is in possession of three different means to express an idea. He named them means of expression:

a. In positive sense

b. In comparative sense

c. In superlative sense

Established means of expression, he proceeds to show how one can operate with them.

1. The initiate can use words understood by all. He changes simply the value of words, depending on the type of intelligence on which he has to act.

For example, the following idea:

The child needs a father and a mother.

Addressing all without distinction of class, the editor will speak positively, saying:

The child needs a father and a mother.

But if he wants to subtract from the understanding of this idea to the people of material intelligence, those designated by the collective term of 'common people', he can express comparatively, from the field of the facts to the law to say:

The Neutral needs a positive and a negative.

The balance requires an active and a passive.

Those skilled in the study of the laws of nature, who are called among us, as a rule, wise men, can perfectly understand the meaning of these laws, incomprehensible to the rustic man and ignorant.

But if it must remove from the knowledge of any truth to these sages, who became theologians and persecutors, the writer goes another step to penetrate the regions of the symbolic. He enters the world region of principles, and then he says:

The Crown needs the wisdom and intelligence.

Accustomed wise to solve the problems subject to him, he realizes, in this case, the meaning of each of the words. But he can not understand the relationship of each other. He can search for a meaning to the phrase, but lacks solid foundation to establish it. He can not be sure of the interpretation. Then, he shrugs. The same does when he finds similar phrases in the Hermetic books. He slams them exclaiming: Trickery and Mysticism!

That was precisely what wanted the editor that the enigma was considered of this way.

2. The initiate may use various signs, as may be the persons to whom he directs.

That was the discretion perfectly employed by the Egyptian priests when they wrote their hieroglyphics or when made use of phonetic or ideographic language, according to the circumstances of each case.

He will see you what is said with new demonstrations. For clarity, the same idea presented above, it is used.

The child needs a father and a mother.

Addressing the crowds, the priest simply design a child placed between his father and his mother or say the verbatim phrase.

If he wants to restrict the number of his readers, he addresses the world of laws and algebraic signs, understood only by the scientist.

Be "∞" sign to indicate the neutral, the representation of the child. Then, we will have:

∞ NEEDS + and -

It can also be expressed thus:

$$(+) + (-) = (\infty)$$

And if he still wants to reduce the fewer people who can understand it, will use ideographic signs corresponding to the principles. Then, he will write astrologically and geometrically:

$$\text{☼} + \text{♂} = \text{♀}$$
$$| + - = \mathbf{X}$$

Soon we will see how these characters, with the privilege to exasperate the curious, do not depend on an arbitrary choice. By contrast, a deep reason presides over his choice.

3. The use of qualitative geometry makes available another method. It is the adoption of a single sign. This can be interpreted in different ways, according as the culture and scope of intelligence of interpreter.

Thus, the ideogram

⊙

It will represent for the ignorant man a circle with a dot in the centre.

The wise sees, the sign is a circle. Its central point, figure astronomically sun and by interpretative extension, the truth. It is rare that cultured man pass at greater depths of the idea.

The initiate sees in this figure the Principle and its development. The idea expressed in his cause, God in eternity.

The described methods have been used to refer to the occult themes of initiation. They are also used in the hermetic books in magic rituals. But, there is also another will, used in ancient times to convey the truths discovered in the sanctuaries. It is symbolic stories.

The best way to perpetuate a thing worth of knowing is by interesting to the imagination, not to memory.

If it is referred a legend to a farmer, it is safe, he will recall it. Passing veiled in evening, they come to posterity the adventures of Venus and Vulcan. It is not so with the laws of science.

It is easy to assume, the case of a villager give enumerating, at the love of the fire, the laws of astronomy. But it should be noted, symbolic stories also contain in their background, not less valuable truths.

The simple man sees in them a nice product of fantasy. The wise discover, with astonishment, the laws of the solar movements. The initiate, by breaking down the names, unravels the key to the "great work". He seizes the three senses, the narrative content (Papus, 1978: 62-65).

With the cited here, it is sufficient to demonstrate the importance of language in the study of magic. It has much to do with this way of encrypting the facts. It is not only found laws governing the story or myth. One must go to its content through the symbolic interpretation of allegory to restore all its semantics. As Papus says: "Do not forget, indeed, that the end we seek is the explanation, though it is very rudimentary, of all those symbols and all those allegorical narratives, which are assumed as shrouded in mystery" (1978: 25).

But this semantic reading system can only be carried out if the pictures of magical correspondences are known. These are as abundant as analogue comparisons can be made. Of course, analogies are not necessarily similarities. By way of example and in correlation with developed before, we can cite the following table taken from Papus:

All facts, however numerous, are classified according to the hierarchy of the Three Worlds. You must add another column for up to two the number of supplementary estimates. They accompany any analogue table (1978: 99).

Suplementaria1: God, according to the Egyptians, the family, 3 stars, the clarity, the elements.

Positive (+): Osiris, father, sun, light, fire.

Negative (-): Iris, mother, moon, moon shadow.

Neutral (∞): Horo, son, mercury, penumbra, air.

Suplementaria2: archetypal world, moral world, material world, material world, material world.

Correlation tables are presented as tables, where each term in bold, leading a column. Each element in the column is sorted. And the third element of a column, for example, may be exchanged by the third element of the other columns, and ritual will remain consistent. These correlation tables are not only necessary to interpret the Hermetic magic symbols and narratives, but also to decipher the magical ideograms.

Generally speaking, one can say, magic was the first human practice in making practical use of linguistics and semiotics, long before they are formulated as science. To encrypt the hermetic knowledge, they come to linguistic procedures for the expression of ideas. When social censorship grew, it moved to semiotic procedures. And no ideas were ciphered, but principles and rules for the magical ritual. When the Inquisition threatened to burn the entire magical heritage recorded in books, the initiate took another step forward. Linguistic and semiotic procedures were merged to facilitate the action of memory. It was not safe store knowledge in writing. It should be the memory of people. The oral tradition of the great ally of magic turned to ensure its preservation for generations. Allegories built with encrypted language and symbols of the great ideas of magic were built then. To popular opinion, they passed through fantastic stories, made to entertain, not to educate. In this way they taunted the social and religious censorship. The magic could survive the ravages of time.

PART FOUR: THE
DISCURSIVE SYSTEM OF MAGIC

Luis Carlos Molina Acevedo

Magic as Information Packet

Another important consideration to study the discourse of magic is the system. The operation of magic is possible only within a system. There are several disciplines from which address the phenomenon. These disciplines have emerged since the implications of the theory of information, such as Dianetics and bioenergetics.

Magic, considered as a system, facilitates the task of understanding the moment when magical discourse becomes packet of information to feed with energy those altered systems (users). Alterations are difficulties in driving and analysis of information. In this aspect seems to be the key to the magic effectiveness through language. But beyond a magical use of language, it is evident ability of language to work on things under certain conditions. It is considered, then, the psycho-social aspect of magical discourse like an information flow.

Information Theory

The interest in this theory, it is addressed to the application of which it has been subjected by the Dianetics and bioenergetics. The latter, particularly, interests from medicine. Dianetics is presented as a new field of research, undertaken from engineering to study the mind. In this development, the best application has been from systems theory. "Dianetics is an organized science of thought, structured, based on axioms defined. It reveals the existence of natural laws with which can be caused or predict uniformly behaviour in the organism or society" (Hubbard, 1977: 79).

The curious thing about this science, it is in the collection of data taken for its development. Mostly, they correspond to the magical arts. Faced with these, the mind takes particular and inexplicable performances from positive science. They are considered by it as superstition. In the background, they cause a physiological effect. It is needed to consider them in the light of the new techniques. "Dianetics provides a basic explanation about the purposes, principles and basis of hypnosis and similar mental phenomena", (Hubbard, 1977: 81).

For Dianetics, seventy percent of diseases are not caused by germs. Aberrations have their origin in the mind. They produce a disruptive action on the circuits. They are something like short circuits. They manifest themselves in disease or physiological dysfunction. They break the balance of the body. The research method of Dianetics is the exploration.

"Throughout my life be travelling from one side to another, I observed many strange things, the witch doctor of the people of Goldi in Manchuria; shamans (witch doctors) North Borneo; witchdoctors of the Sioux; cults of Los Angeles, California, and modern psychology. Among the people who were questioned about the existence, it was a magician whose ancestors had served at the court of Kublai Khan and a Hindu who could hypnotize cats. Raids were made in mysticism; information was studied from mythology to spiritualism; a little here and there, a bit of everything" (Hubbard, 1977: 14).

From this perspective, the lowest common denominator of all existence is to survive. "The basic command information on operating the body and the brain was Survive! That was all; there was nothing out of it" (Hubbard, 1977: 19). But in the mind of man, it had also demons.

The brain was seen as composed of circuits (see picture 1 at the end of this book). The optimal brain was a simple circuit. To this, circuits "demon" were added. By ordinary electronic, it could install all kinds of "demons" observed. Thus, it became clear there were no demons or ghosts, vampires or Tohs, but there are aberrant circuits (Hubbard, 1985: 25).

The image 1 is a block diagram. A controlled system (S) and a system controller (R) are observed. It works by taking the output (Y). This output compared with a standard (Z). Thus, it is known whether the changes made to the input (X), correspond to the standard or not. If the result does not fit the standard, the feedback mechanism operates to feed back the system again with a modification (Δx). This modification may be amplification if (Y) is less than (Z). And it will be a reduction, in the opposite case.

The optimal brain was a brain without aberrations and therefore, it was the basic personality. This led to cover the issue of memory. It was necessary to reconsider the concepts of unconscious and conscious. By Dianetic sleep, it could lead people to remember the forgotten, even from the moment of conception. "Let's call the 'unconsciousness' with a new name: Anaten (analytical attenuation), attenuated analyzer" (Hubbard, 1985: 54). The Anaten is a reduction or weakening of analytical consciousness of an individual. Therefore, the difference between man and animal is the existence of an analyzer in mind. "Demons" occlude it or falsify, but without altering its database. Inaccuracies caused by electric shock or injury are not discarded. In this sense one could speak of a magical efficiency from the reorganization of information. This allows the magical use of language to strengthen again the attenuated analyzer by the aberration.

"The function of the mind included to avoid pain. The pain was anti-survival. Avoid it!" (Hubbard, 1977: 48). This function is entrusted to the analyzer, which prevents the entry of the irrationalities of the outside world. That is, try to evade the aberrant perceptions. So, while the animal uses only identities, man has an analytical mind. In addition to the reactive mind, man thinks in differences.

Magic could look in the light of these new findings as a confirmation of the great aspiration of Giordano Bruno; the magical memory as a reform of the mind. The magic, through language, seeks to introduce information of identical type, irrational and absurd; so, where perceptions are equal to all the words, like objects, like the painful situation of the past, already lived. Now, by a partial identification, by reconstituting from a set element, it is revived or updated that situation painful to an in now-here.

Much of the alterations in the functioning of the body of man, they are originated in equalling everything with all and of a part of the thing with the entire thing. Apparently, a large flow of these aberrant elements are represented by words stored in engrams. These, faced in the pronunciation of the keyword, are taken to the representation of the original situation. An engram is an image of energy. It is formed during a period of physical pain, when the analyzer is out of circuit (due to fatigue or pathogen state). The organism experiences something contrary to survival. It is not matter if is by a belief, or something actually real. "An engram is acquired only in the absence of analytical power", (Hubbard, 1977: 54). In light of this, it is not so farfetched the effectiveness of hexes. Simply, it utters the appropriate word to trigger the engram already stored in the mind of the person or victim. It is also easier to understand the effectiveness of magic in its effort to restore the analyzer to return the deflection circuits, to return the balance to the organism as a variable, operated within certain limits.

The magic operates with language on engrams. Aberrations expressed by the user, they are disintegrated by a loss of intensity on engrams. These are archived or deleted from memory. Thus, the magic, in most cases, stops the effects, the function of engram to return the organism to functional equilibrium. Only in a few cases, it gets to eradicate finally the aberration or engram. So, alteration may appear again over time, as the only engram was filed, not eliminated. This allowed again the normal flow of information required for proper functioning. However, it should not be thought that magic and Dianetics are omnipotent, they have also their limits.

Dianetics can break the habits simply by emptying the engrams. "Dianetics could only change a training pattern (placement and arrangement of certain automatic responses to various situations and mechanical action by the analyzer), if the individual would consent to it" (Hubbard, 1977: 62).

Magic System

The above considerations should be made clear why it is necessary to see magic as a system. In addition, decipher how it circulates language as information (see image 2 at the end of this book, for a better understanding of the following discussion).

The universal graphic of language can be established between the poles of change and structure. The change includes the categories of process and speech, that is, the diachrony. The structure includes the ones of system and language, that is, the synchrony. The intersection of these categories is the word. It links diachrony and synchrony, to the speech with the language through the discourse and the process with the system through the event, so as the event and the speech between them (Fuentes, 1972: 33).

In the magical ritual, it is restored a moment of foundation through language, through discourse. The gods made possible the world, they created it. To the man, as a creating at the image and likeness of them, accounts the foundation of world by the implicit segmentation in his language, by taking a symbolic possession. This installs him in this world. If the myth recalls the creation of the world, magic updates its foundation. Therefore, despite the fixity of magical language, its essence is in the staging. The magical action is constantly changing due to new contexts where it is updated.

For Levi-Strauss (1986: 25), "everything happens as if music and mythology did not have need of time more than to give a lie. Indeed, both are machines for eliminating the time". But magic, however, is a return for founding the original time, and with it, the world. That is, the time is not stopped for narrative recreation, as it is in the myth, but by a permanent creation of the world from its re-segmentation within the language and symbolism, for a magical operation.

Image 2 represents the diagram of language (see image 2 at the end of this book). The language diagram gives us an overview of the staging of magic. The magical language reverses structures to the world of change, where the time of the foundation was not. It is happening now by a joint of the central and eccentric instances. It is a discourse based on historical facts. It enters into an analogue process to become synchronous by the system and language, objectified in a series of magical formulas. The magical discourse becomes event by word, by the enunciation of the magical texts. By language, magic becomes universal with a centre in everywhere and nowhere at once.

The magic is universalized by the resulting tension between cultural credit side and technological debt. It globalizes by tension between the expression of magic in each community and the expression of magic as a whole. For them, it is possible to criticize the social reality as cause of dissatisfaction for the user and magician.

Luis Carlos Molina Acevedo

Black Box

The most problematic concept of systems theory is the one of the Black Box. "This procedure consists in prescind from the detailed study of the system and interpret its behaviour from the relations of inputs and outputs of a transformation, assuming that the input-output is stable enough to make a prediction" (Rotundo, 1985: 73). But in the magic the primary relationship between elements of the system is the one of magician with the user. This relationship can not be predictable. It involves people. It needs to be addressed within such a Black Box. It takes in account its internal processes.

The magic is staged by the ritual, and as such, share certain characteristics with other activities based on the ritual. Interdisciplinary research has allowed observing what happens during the ritual (for a better understanding of what explained here, see picture 3 at the end of this book). An intensification of emotion expressed by certain physiological reactions is given.

There are three components of emotional process:

1. Physiological

2. Cognitive

3. Regulatory

The first is mediated by the second and third. The data provided by the biochemistry, confirm the existence of the relationship between cognitive processes and emotional states. Indeed, the cortical brain centres, based on the valuation of perceptual data sent to the hypothalamus neuro-chemical information. It secretes hormones to stimulate by vascular pathway the pituitary. This produces adrenocorticotropic hormone (ACTH). When poured blood, it stimulates the adrenocortical activity. The cortex of the adrenal gland increases production of corticosteroids (cortisolo). At the same time the hypothalamus induced by stimulation through sympathetic, the production of catecholamines (adrenaline and noradrenaline) in the adrenal medulla. The effects of increasing the rate of these hormones in the blood are:

1. Dilation of the pupil

2. Tachycardia

3. Vasocontriction

4. Increased blood pressure

5. Shuddering

6. Sweating

The last two induced by the hypothalamus through the sympathetic without intervention of the adrenal glands. All these phenomena can be defined as "Increased emotional reactivity" (Scarduelli, 1988: 82-83).

This physiological process is complemented by the already approached from Dianetics. Scarduelli insists point, the secretion of the adrenal glands not part of emotional stimuli. It is originated in cognitive processes. In other words, it is possible through language the magician brings the user to an endocrine process with physiological effects. The following table, offered by the author, gives enough elements to show how the information acts or magical ritual with its language, in a reordering of the information for the elimination of altered states.

The image 3 shows the pattern of physiological reaction of people in situations of everyday life (see picture 3 at the end of this book, for a better understanding of what follows).

As shown in the scheme, on the basis of cognitive development and valuation of perceptual data, cortical centres send information to the hypothalamus. This translates into orders transmitted by nerve pathway, to the terminal fibbers of the sympathetic and the medulla of the adrenal glands and, by vascular pathway, to the cortex of the adrenal sympathetic. Production of catecholamines and corticosteroids by the adrenal determines a series of biochemical changes. They constitute the physiological basis of state of emotional arousal:

Sweating

Horripilation

Vasoconstriction (paleness)

Tachycardia (palpitations)

Paleness may be a sign of a great fear or an uncontrollable rage. Palpitations may be a prelude to a terrified leak or a burst of joy. One or the other depends on the valuation of perceptual data. That is, a cognitive process culturally mediated (Scarduelli, 1988: 84).

It should be noted in the above, the way as information is conveyed to the analytical mind and reactive mind. Similarly, bioenergetics, in an interdisciplinary study with medicine, has also found information flows. In the healing acts there is a sending of information via medication, via laser beam, via needles. That is, it does not relieve the drug itself, but the information circulated by systems of conduction of the information until the circuit where the alteration is. In this regard, they are very interesting results obtained with placebos. The most interesting of these investigations is the identification of seven systems of conduction of information into the human body.

A global system, of vital energy regulation, should include the following control systems:

1. Nervous system: a. central, b. autonomous.

2. Humoral systems: a. endocrine, b. indoletial reticulum, c. diffuse neuroendocrine or system of neurohumoral integration.

3. Ubicuitario system of peschinger or environment-cell system.

4. Ubicuitario System of bio-plasma with solid semiconductor, that is, macromolecules with piezoelectric properties and semi-conductive.

5. Ubicuitario System of the biological micro-organizer: melanin, Neuromelanin, DNA (with possible superconducting property).

6. Ubicuitario system of global integration: bioplasmic body or etheric field and his corresponding systems of signal translating (chakras of the Hindu worldview).

7. Ubicuitario system of cosmic integration: noosphere (from view point of Teilhard de Chardin) or field of transpersonal integration (Carvajal, 1986: 38).

Then, it is has that the process of internal conduction of information in men is complex. With each advance, it is more understandable the effectiveness of magic. It is like the magician, through his cultural education in a community, learns to do some things to produce other. The art of the magician is to master the mediated action from a ritual mediated by language and symbolism to trigger a series of biochemical processes. These, in turn, lead to physiological effects. He, unwittingly, applies a whole theory of information, because as shown by the theory of relativity, energy and mass are equivalent and transmutable ($E = mc^2$). The mass is energy and energy is information, from which is concluded that all is information. The information can only be modified by information. Information with new data, changes to the information of the old data.

The magician takes the symptoms presented by the user and translate them into a magical semiotics, it sends information to the body of the user to produce certain reactions. Thus, he obtains a possible correction of the alteration. This would be the essence of magical efficacy.

The ritual use of cadence chants, psalms, processions, drumbeats or rattles, i.e., repetitive movements and visual and auditory stimuli, visual and auditory stimuli, rhythmic, causes a hyperstimulation of the sympathetic. When a saturation level is reached, an overflow phenomenon on parasympathetic is done. This induced stimulation (accompanied by phenomena of 'resonance' in the two cerebral hemispheres) causes perceptions of ecstatic type. In particular, a sense of togetherness and overcoming of opposition is experienced. It is of variable duration but generally includes most of the participants. It assumes different aspects depending on the nature of the rite (Scarduelli, 1988: 84):

1. Merger with a higher power

2. Disappearance of the fear of death

3. Feeling of universal harmony

The Interaction

For Leach (1966: 242), the "typical" rite is usually regarded as a long and complex celebration. It is a series of acts endowed with a characteristic structure: A divisible "discourse" in "paragraphs", "phrases", "words", "syllables" and "phonics". Indeed, if a rite is a communication process, it must be based on a code. "The whole ritual acts should be articulated according to a coordinating logic of the basic units of the code", (Scarduelli 1988: 54).

Magic, as an activity of the imagination, goes to the ritual. So, it should be seen as a communication process. However, as the ritual is repetitive, seen from the information theory, it would be a redundant emission. But seen as a communication process, each redundancy is significant, can not be eliminated.

The analysis of human rite is infinitely more complex compared to animals. The rite is not only structured according to the coordinating logic of a code comparable to linguistic. It contains also verbal enunciates (formulas, invocations, comments, cantos). Therefore, the linguistic code is, at the same time, the general model of the structure of the rite and one of its components. The overall result is the structure of a number of different codes (language, gesture, and colour) (1988 Scarduelli. 54).

The communication process of a magical ritual, it is characterized because communicative acts are essentially made by the magician and observed by the user. Linguistic communication is displayed as the starting time and contributes to clarify certain steps or procedures.

The decoding of the message, by the receptor, is based on knowledge of one or more symbolic codes belonging to the cognitive system shared by all participants in the rite. Decoding occurs through the recognition of the symbolic values associated with a number of heterogeneous elements:

1. The physical space in which is developed the ritual action

2. The timing of the execution (hour, day, month, season)

3. The objects used by actors (knives for sacrifice, relics, images of deities, magic potions)

4. Elements of diverse nature that individualize the social identity of the actors of the rite (sceptres, insignias, body painting, ornaments, badges, masks)

Communicative action is governed by a structure of different codes within the magic ritual. It includes also a symbolic element. "This network of symbolic meanings, which is the context of the ritual action, helping to found the appropriation of the reproductive capacities by men and legitimize their supremacy, which is also stated in the execution of the ceremony, by defining of rolls" (Scarduelli, 1988: 66). Emotional excitation corresponds to a structure of communication in the magical ritual. This ritual action can sort the mindsets of the user. At the same time, there is also an orientation of social behaviour. Sometimes, it manages provoking emotional and physiological reactions.

Interaction, marked by the presence of the user, allows activate the magical system. It is established to a state of closed like if it was a fragment of world history. User symptoms are collected into a process of translation by the magician to make rearrangement of the information in accordance with the model of the world of community where the ceremony takes place. In this way, the user to receive it, it is given an action on the disturbance of him. His body turns a steady state. Thus, the magician reintegrates the user to social group again, it makes him functional.

Luis Carlos Molina Acevedo

Case Analysis

A story about a cure will be taken to see how the magical system works. The polyphony of voices will be the basis for significant reconstruction. That is, from fragments of the phenomenon expressed by different informants, it seeks to restore the overall process. It should also be clarified that the names are fictional to protect informants.

Dona Maria lives in the town of Fredonia, Colombia. She relates, once one of his sisters was dying. A neighbour, after seeing her, said: it had made her sorcery. The best thing was to go to Don Pedro, who lived in the neighbourhood Antioquia in Medellin. To he should just send him a lock of hair and some urine of the patient in a jar and with this he knows if could cure her or not. Don Pedro to see the samples, said to the children of Dona Maria: I can cure her. He took two bottles filled with clear water. He explained how the remedy should be given to her. He promised, as soon she was recovered, he visits her. They were given liquids to her. They tasted to water, without more. They were accompanied with drinks of herb of grace at night and basil in the morning. Fortnight after she threw the first slug. She had it in the gut. Another neighbour woman, who was present there, ran to throw the slug to toilet. The patient returned to worsen. Dona Maria sent her children to where Don Pedro again.

He asked the details. Then, he said: relapse was due to the failure to burn the slug. The second slug was thrown fortnight later. When this mollusc was burned, it jumped like a balloon. The patient improved considerably, while the person causing sorcery (a mistress of husband) sickened. When she threw the third slug and it was burned, the mistress died. The patient was cured. Then, Don Pedro came home of sick woman.

Dona Maria explains, the neighbour woman who rushed to throw slug to toilet, was a friend of the mistress. She made sorcery to her sister to get her husband. The mistress died because evil was returned towards her. Here, it is necessary an additional voice from other informant. The first neighbour became clear the sorcery. The question is how she knew it. The answer stands present the semiotics of sorcery. She knew it because she saw that the sick woman took several days without speaking. She showed sad and melancholy. Still, doctors were not able to discover the bad or the cause for her condition. With these symptoms, she took a first diagnosis, which must be confirmed by a healer-magician.

There is an initial translation of the condition of the patient to a semiotics of magical type: the sorcery. This action was sufficient to activate the magical system of the heal-magical-action. In turn, the healer-magician translates a natural signs to a magical semiotics called sorcery. With this, he comes to handling information. He gets the item or product to introduce into the user the sufficient and appropriate information. He rearranges the altered information that is source of imbalance in the organic system of the user. Much of the processing of information, by the magician, was concentrated in the liquid of the two bottles. Other minor part, it was concentrated in the orisons done at distance. The magician uttered these orisons to carry the sick to recover. When the patient swallowed the liquid, this, as vector, directed the information toward altering for stopping the physiological effects. That is, an information type was evicted to accommodate a new one. The performance required to return to the equilibrium position of the system was restored. Such eviction was symbolized by the expulsion of slugs. In other words, semiotics, called sorcery, took shape in a slug. It was thrown and, as a result, came the relief.

The Engram, if it had not been destroyed, would have continued to produce changes from time to time, in the information system of the body of the patient. Therefore, it was necessary to burn it in its objectification, to thereby act symbolically, on the bank of red signals. This was in a process of identity when it should be a process of differentiation. In this, it was consisted the sorcery. "What is tuberculosis? It is a predisposition of respiratory system to infection. What is this? What is that? Now you have the proposition. It works. Psychosomatic disorders, arthritis, impotence, this or that disappear when engrams are removed from the bottom" (Hubbard, 1977: 74). It is amazing this new view of the disease. Most of these engrams are cultural products. Sorceries would not exist, if the culture of a given community not creates the willingness of its members toward them. Similarly, it creates social mechanisms to eradicate them. Remember, the basic principle of the disease is the one of organic rejection of certain types of operation, and such rejection is cultural.

The action of the ritual of burning slugs, it brings into operation the analyzer. The reactive mind ceases to set wrong and abhorrent identities between words and painful situations bygone. These are culturally induced by the community. They are stored in the worldview of the user. The main component is the magical discourse transmitted by oral tradition. This is full of sorceries, herbal-magical-actions, evil eyes, incantations and other magical semiotics of alteration. The user, from the identification of one of the elements, trigged the semiotics of sorcery. She reconstituted such semiotic in her body. That is, the sister of Dona Maria, by the identification of a word or gesture or situation in her state of deceived wife, restituted the semiotics of sorcery culturally stored in her mind. With this information, her body adopted the corresponding operation. Then, Don Pedro enters in scene. By a rearrangement of information, he returned to the body of the sick woman to functioning socially accepted. By reactivating the red data bank, it was returned the engram. Therefore, it was precise to nip in the bud it.

Only from the acceptance of the above, it would be explicable the reversal of the engram over who creates it or reactive externally. The sorcery is essentially an external act to the patient. For it, a partial element is introduced in body of user. This is able to lead to user to restore the engram culturally stored by the victim. In magic, it operates essentially with information over information. When it does damage, it does on aberrant way for the mind. But it can do it on des-aberrant way for healing.

Different case occurs with the analyzer. This establishes a pattern of consensual behaviour. It helps the bank of dangerous experiences to establish a series of reactions to actual danger signals. It does not react to engrams, staged or represented in a ritual. However, it is not dispatching these considerations lightly. It is not simple belief or superstition in the way of positive science. The task is to read further to clarify why man has needed these cultural configurations. They become the primary means of survival. They include magical elements for having an abhorrent mind or not abhorrent mind.

Continuing the story, it is necessary to introduce a new voice to clarify certain aspects of this case study. Rosa is a practitioner of black magic. She says: to introduce sorcery on a person, such as toads in the stomach, one proceeds to draw with coal such an animal in a foil. She clarifies, the precision is not required in the layout, the important thing it is the gesture. This is done after twelve o'clock. The orison of Black Angel is prayed where the sorcery is dedicated to the victim. Since then, the animal begins to develop into the stomach of that person.

This information allows the operation to restore mistress. Surely, she performed a ceremony similar to that described. Somehow, the existence of this ritual came to sick woman. She restored the semiotics culturally acquired to become it an engram. Her analyzer circuit was partially out by a state of distress.

It was produced by the knowledge about husband's infidelity. Her husband had a mistress. There is a weakening of her critical mind. "A lock is a state of mental anguish. It depends for having power of the engram to which it is attached. The lock is more or less known by the analyzer. It is a time of strong restimulation of an engram", (Hubbard, 1977: 70).

So far, everything is understandable, but to understand the alteration reversal with fulminant effectiveness, we must introduce another voice in the story. The second neighbour woman threw the slug to the toilet. She is allied with the mistress. She is the conduit to reverse the bad. She operated the information back to the starting point. She must have communicated to the mistress the existence of a powerful healer-magician. He had taken out a slug to the sick woman. Thus, information, objectified in a slug through language, could act against her more strongly. Sorcery was no mediated by translation of a language, but as energy immediate taken from herself. It was his creation sent back toward her. This creation, joined with the belief from magician in power differentials, had more effectiveness. A more powerful magician can break or interfere with the work of another and even to send it back. So, it is important the role assigned by the magician to the possibility of the existence of enemy spirits. Therefore, in their invocations and rituals, he provided by the use of symbolic objects for his defence.

It should be made again record at this point, about the purpose of this study. It does not seek the truth of magic, but the interpretation of magic as discursive scenario where social reality is criticized. Sometimes, it is presented as something difficult to live by the rigidity of its conventions. This is the case of this patient. She not conceives the fact of being deceived. Her husband has a mistress with consequent social censure involved for her. The social group may think that she is not woman enough to make her husband happy. So, he should look for other women. Hence, her mental anguish. It weakens her analyzer. She takes as the real, a dramatization. It causes only dissatisfaction. But, she can take the help of a magician to kill in the theatre to the rival, without facing social sanctions or legal processes of accountability.

In this sense, it is not so important if that story is true or not. The important, here, is the reconstruction of the scenery. This allows us to appreciate the intricate workings of magic as an activity of the imagination. It turns to fiction to effectively correct changes in the real. This should be understood, in turn, as a construction of world socially accepted, as a model of the world. By representation, it is continually modified. The idea of it, into every individual, is set permanently.

This story shows a wife in a world where infidelity is considered bad. If the husband has a different woman, then, they are socially justified certain actions. These are operated in a world where many things can be arranged by art of magic. Especially, where there is a clear conviction, language can be objectified in things and animals. These can also be introduced into people by symbolic procedures. Simply, it is needed the representation of that animal in foil and then it is restored into someone's stomach by praying some orisons. All this is integrated into a coherent discourse. From this derives its effectiveness.

The symbolic structure of ritual practices allows to participants to make an aesthetic distance from their emotions. That is, a state of balance between affection and thought. This balance allows maintain a deep emotional resonance within the limits of cognitive control (Scarduelli, 1988: 88).

This means, man always has maintained spaces to experience aesthetics as a life experience. When life is not yet posed as an aesthetic experience, like in postmodernism, people already had a cultural place where contact with the experience of imagination, beyond the classic or elitist art. That place was the magic ritual where the staged it is really lived. It is not experienced as a theatrical artifice. So, the objectification of language in symbols is very important. Here, the support material is life itself. It is not frames or tables of representational art to define its scope. Here, the main raw material is language to paint, to represent toads, tadpoles, slugs and a thousand more of sorceries. They are the confirmation of the existence of the afterlife as something wonderful. It becomes a place where one can go when one try to put meaning back into everyday life, the be-here. The communication process, of magical ritual, allows the staging of an aesthetic experience of life. It is free of the restrictions and limitations of the social. This is not possible for art, or other practice socially accepted.

The expressive process, where the ritual is articulated, produces cathartic effects. It resolves at the individual level social contradictions. It is not because it makes free individuals from alleged existential angst. This second objective is achieved by the idealized representation of society. But the first objective, it forces individuals to break free from the tensions repressed. They accumulate in everyday life through communicative modalities. They interpret again the behaviour to make it compatible with norms and values socially sanctioned (Scarduelli, 1988: 88).

By this staging by communication, the limits of compatibility between sociocultural coercions and interpersonal adaptation are redefined. Thus, the wife can justify that her husband has gone for another woman. She could not do anything about it. She was the victim of supernatural forces. But at the same time, it is a conviction herself. She can keep her self-esteem. Her husband did not betray her with another woman. He was also a victim of these supernatural forces. To be neutralized, the institution of marriage is restored socially. It is not free that the main symptoms of sorcery are the breaking of interaction, the social isolation by self-absorption.

Symbolic Reading

The above considerations, about the case analysis, require a supplemental reading at symbolic level. It shows the whole story revolves around a symbol. It is, therefore, the most important: the slug. The slug as a symbol is objectification of language under the name of sorcery. In ordinary language it designates also slug to the mollusc or "mother" of garden snail. The people identify two kinds of slugs, one with shell and one without it. The slug of this story acquires a special isomorphism. It can explain better why the sorcery is objectified in a slug rather than a frog or tadpole, as it is the usual. One can not say, this fact corresponds to an individual style. Dona Maria, thereupon, related other case, the one of the mother who in Bogotá had been removed a number of tadpoles from her stomach. It is also popular belief that the slug has not shell because it is sloughing home or shell. They say that it abandoned its shell to find another.

The story establishes a link of isomorphism between the slug and the spiral as a symbol of time and progress. This is linked to the constellation or the archetype of the wheel representing the rhythmic pattern. With this as a social alteration finds support based on the sexual. The rhythmic pattern corresponds to the dominant reflection of rhythm. This governs the laws of reproduction, of becoming, the myth of the eternal return. The spiral of snail is progress to infinity. But, you do not forget that this isomorphism is the objectification of some sorcery. Next to it, it is grouped a series of symbols located within the night regime of image. The symbolism of the spiral usually belongs to daytime regime of the image, but, by a euphemism, it has gone the opposite regime. There is an inversion of value assigned to the terms of the antithesis. Keys to this are found in the scheme of descent, followed by the objectification of sorcery, of the slug. This is thrown from the intestine where it was housed. The spiral is contaminated by this single anthropological haul. It becomes to share an isomorphism corresponding to nocturnal water. Slugs are defecated. The "mother" of the snail, the mother of time and progress is given birth by a reverse conduct, and as mother, takes also a night isomorphism. Then, it is had a total euphemism about the faces of time to outwit its inevitable march.

The complementary symbolism to it is represented by pure water, clear water. Therefore, it is located on the daytime regime. Dona Maria was surprised to find that the colour water tasted like water. It was not contaminated or mixed. This puts the healer-magician within the daytime regime of image, who fights against the night regime of image symbolized by slugs. Thus, the heal-magical-action is presented as a struggle between the two regimes. He does to feel, to the mistress, her defeat. Consequent death comes as a triumph of good over evil.

Magic: Symbols and Texts of Magic

From a symbolic perspective, it is not hidden in the story a reversion against mistress. It is not punished in it an attempt to outwit the inevitable flight of time, but its intention to annihilate time. At reversing the face of time, the mistress objectifies in the slug the time own as a parasite. This should suck the time to the victim to bring her to death. At being burned the symbol, it was destroyed the time of magical woman, that is, her life. The slug was something like a noise in the vital information of the user.

However, please note, the choice, of this symbolism for the expression of the story, is unconsciously motivated by the sexual as dominant reflex of the rhythmic. By sex man seeks to overcome the inexorable passage of time with his stay in the offspring. All of these actions to provoke the sorcery, they are based on the sexual. A mistress intends to stay with the husband of another woman, who by her very social condition, can go to the social function of magic. The magician accepts risk to help her. Evil may revert to him by helping her. But socially he is authorized to do so. This authorization is given by the institution of marriage. The wife acquired the right socially. She has the right to defend her marriage. The magician can operate without fear of social sanction, or the consequences of magical action itself. Thus, healing-magical-action is a triumph of white sex, socially permitted over the black sex, forbidden. White magic trumps over black magic projected into sorcery as objectification of a language called "Orison of Black Angel".

Luis Carlos Molina Acevedo

Social Reading

The semiotics of sorcery has also a social reading in practice, made by members of the community from a cultural competence. When people are questioned about how they know if a person has sorcery or not, immediately responds: Because they are as sad and not talk to anyone, seeking solitude, walk with his head down. In conclusion, people with sorcery go from being socially normal, to become sad in a hasty manner. Communication and social interaction are altered. The social reading begins with an acknowledgment of the discontinuity in the way people behave without apparent cause. This is interpreted as a semiotic of sorcery.

It should be noted, in places under observation of the magical phenomenon, here presented, occur the opposite to the one described by Lévi-Strauss about the symbolic efficacy of magic. There, the sorcery does not take effective by marginalizing socially the victim. On the contrary, some special solidarity for her it is waked. Family shows more interest. It worries about finding who cure her. This means, so the magic is of all times and peoples, it is possible to find differences in its application and understanding between communities. A different valuation is made of it. Thus, in this region of Antioquia, despite the great fear of black magic, there is also conviction that good always triumphs over evil.

The main social symptom of sorcery or any alteration of magical type is in communication breakdown due to a mismatch in driving information at the somatic level. Therefore, it should be scrutinized the black box, if one wants to see the magical action in its true psycho-social dimension. The same process happens in reverse. When the sorcery is done, an alteration occurs in the external signals of body. It does not analyze or leads properly. It is altered or the conduction of information is interrupted. "The problem of disease is basically the problem of self-recognition", (Carvajal, 1986: 34). Or if you want, "the disease is simply the result of an alteration in the information, regardless of the name given to it." (Payan de la Roche, 1986: 10).

According to the famous definition of Goodenought (1964): "... The culture of a society is all the one there is to know or believe for acting in a manner acceptable in the face of its members and to do so in any role they accept for themselves". Culture is the entire one it should be known. Part of this knowledge is the conversational competence. "The general view we have of culture, it is the system of interlaced signification, and a fundamental idea, in this sense, is that you can not be incommunicado: even if you look impassively to anywhere, it is communicating" (Stubbs, 1987: 23).

Characterization of Magic Act

According to the above considerations, and in order to differentiate the magic of fraud, one can enter to see the magical act as an action. It can be characterized by the method used by Searle (1990) for the regulation of the illocutionary act of promising, as follows:

As a **magician** M **emits** a **healing text** HT in the presence of a **user** U, then, at to emit literally HT, M proposes sincerely **to cure** C to U if and only if the following conditions are met.

1) Normal input and output conditions are given.

This is a condition for the interaction between magician-user. The **output** covers the conditions to speak and gesticulate intelligibly. The **input** covers the conditions of understanding of verbal and nonverbal language of magical ritual. Both include things like:

a. Magician and user are immersed in a culture where there is the magical text

b. Both are aware of what they do

c. They have no physical or moral impediments to the magical interaction

d. At no time they consider the magical ritual as a theatrical performance

e. Others discourses are outside, if they obstruct the magical interaction.

2) M expresses the propositional content C at to emit HT.

This condition isolates the healing propositional content of the rest of the magic act, understood the healing text composed of a propositional **content** and a **force** of enunciation.

3) By expressing HT, M enunciates a healing power HP of M.

In a healing act must be done an act of the magician. This act can not be in the past, but in the present with future repercussions. The healing act includes series of acts. It may also include states and conditions. To that extent, the conditions number two and three are the propositional content. Therefore, this condition can be reformulated, because about the objects are pronounced different expressions, as follows:

At to express HT, M pronounces an expression of power of M, whose meaning is such that if the expression is true from magician, it is true that the magician has HP.

4) U prefers that M applies HP to that not to apply HP, and M believes that U prefers that he applies HP to that not applies HP.

This would be the crucial distinction between curing and make spells as magic acts. While a cure is to apply power to the user, the sorcery is to apply power to an enemy of the magician or user. In that sense, healing must have the consent of the user. A special situation where the user wants to cure is required. When the healer-magician fulfils a social function, the mere fact of a person go to him, it implies consent from this person. The magician is conscious that the person is looking for his help. In the case of sorcery, this is set against the will of the person. The situation, in this case, is given by envy or revenge, the desire to do evil. It can be said then that the concepts of good and evil are determined by the presence or absence of consent. This basic foundation determines the classification of magic in good or bad. The good is mediated by consent, the bad is mediated by the absence of consent.

5) It is not obvious to M or for U, that M will apply HP in the normal course of events.

When the user goes where the healer-magician, obviously, the user is not doing or will do the things necessary for healing. He needs someone who tells him the procedures. It would be out of place for user to go where the healer-magician, if he knew that this is going to cure him without asking. Thus, conditions number four and five act as **preparatory conditions**.

6) M intends to apply HP.

The difference between a magician and a charlatan is in the intentionality. While the magician has a strong conviction to cure his user, the charlatan just thinks about the money from his client. When the magician party for action of his intention to cure the user, entails, he thinks can do or refrain from doing so. To that extent, this is a **condition of sincerity**. This condition is not present in the charlatan.

7) M tries the emitting of HT will place him under an obligation to apply HP.

The essential feature of stating a healing discourse is to apply a certain power. In that sense all magical acts are related. This would be the essential condition of magic and its enunciation. Proof of this can be the gesture of respondents of offing his hat when going to express a magical text in front of the recorder of the researcher. It is not through ritual and however there is a great respect towards the magical text. This has great power, although it is not being used to make magic.

8) M **tries** (T) to produce in U **knowledge** (K) over the emission of HT as the fact of placing to M under the obligation to apply HP. M tries to produce K through knowledge of T, and intends that T be recognized by (through) the knowledge that U has the meaning of HT.

The user is interested, above all, in the power. That part of the sense of magical text does not circulate by the language. Remember, in all cases, the user perceives only a whisper. He does not listen to the words as such. In this whisper, he attributed all the magical sense of the text.

This shows what for the magician is to want say his emission with the meaning of a healing. The magician intends to apply some healing power. With his actions, he makes the user recognize his intention to apply that power. He intends also to achieve that recognition, under the meaning of the healing text. His emission should be associated conventionally with the application of that power. But in no case can be said, the effect of such an act is the product of a convention. In this case, the magician supposes, semantic rules (to determine the meaning), of the emitted expressions, are sufficient to embody a power, a force. The rules, to put it briefly, make it possible, by making the emission that the object of the intention expressed in the essential condition number seven is achieved. And the articulation of that achievement, the way as the magician gets to carry out his task, it is described in the condition number eight.

9) The semantic rules of dialect spoken by M and U are such that HT is issued correctly and truthfully if and only if are fulfilled the conditions number one to eight.

This condition number nine aims to clarify feature of the healing text emitted. This is such under the semantic rules of the language. It is used to make a cure. The meaning of a healing text is completely determined by the meaning of its lexical, syntactic and symbolic elements. This is precisely another way of saying that the rules of emission are determined by the rules applied to its elements. Then, the governing rules of the item or items shall be made. They serve to identify the enunciation force as a cure.

The condition number one is interpreted in way broadly enough. Along with the other conditions, must ensure U understands the emission. Together with conditions number two to nine involved, the enunciation effect K occurs in U through recognition by U of the intention that M has of producing it. This recognition is achieved by virtue of the knowledge that U has about the meaning of HT. This condition always could be enunciated as a separate condition.

The next task is to get out of the set of conditions, a set of rules for the use of the indicator device of declarative force. Obviously, not all conditions are equally relevant to this task. The condition number one and the conditions of forms the ones number eight and nine apply generally to all genres of normal magical acts. They are not peculiar to the act of healing. The rules of the indicator device of declarative force to cure correspond, as will be seen, to conditions number two to seven.

The semantic rules for the use of any **indicator device of declarative force to cure** FC (force to cure) are:

Rule 1. FC must be emitted only in the context of a healing text HT, whose emission predicates some healing power HP of magician M. This would be a rule of propositional content. It is derived from the conditions of propositional content number two and three.

Rule 2. FC must be emitted only if the user U prefers that M applies HP to that not to apply HP and M believes that U prefers that M applies HP in front of that not applies HP.

Rule 3. FC must only be emitted if it is obvious to both M and U that M not applies HP in the normal course of events. The rules number two and three are preparatory rules. They derived from preparatory conditions number four and five.

Rule 4. FC must be emitted only if M intends to apply HP. This is the rule of sincerity. It is derived from the condition of sincerity number six.

Rule 5. The emission of FC accounts as the assumption of an obligation to apply HP. This is the essential rule.

These rules are ordered: the rules number two to five are applied only if the rule number one is satisfied and rule number four applies only if the rules number two and three are also satisfied (Searle, 1990: 65-71).

With this exploration of the magical system has been evidenced that in the magic is given the expression of an act of different enunciation. This seems not to have been considered into the theory of the speech acts. At first glance, it appears as an act to other with common features. It shares features of illocutionary acts such as of ordering, of promising, of asserting, of ordering and even of threatening in cases of sorcery. It shares traits of perlocutionary acts such as of persuading, of convincing, and, why not, of suggesting. But obviously, is not one of them. Necessarily this kind of act of enunciation, own of language in the magical action, would be the one of **PERATING**.

Traces for this suspicion are found in expressions of ordinary language such as "to operate miracles", "to operate wonder things", "to operate prodigies", "to operate impossible things". That is, in these expressions the verb **OPERATE**, refers not to operate in the normal course of things, but another to operate. It takes means no obvious to men in general. However, it manifests through the facts, they show its effects. But the main trace emerges from reading of the following passage:

"Although the definition of magic is presented in a hasty manner in different writers, it can not be denied that there is in all the same fundamental thought. In all ages and in all countries has been fuelled the belief that apart from the normal way as changes in the world occur, through causal relationships of the bodies to each other, there must be another completely different mode that is not based on these causal relationships. The means employed in the latter seem, therefore, manifestly absurd, considered from the point of view that characterizes the first mode of action, since the disproportion exists between the causes immediately in sight, that any causal between the one and the other was impossible. It was necessary to assume that, in addition to the outer link, which established a physics nexus between the phenomena of this world, could have another that had its beginning in the being itself of all things: a link, so to speak, underground by which was established a metaphysics nexus from one point to another, and immediate action could be produced. It was necessary to suppose and admit that one could operate over things in his "inside" instead of, as usual, to operate in his "outside"; It was necessary to admit that the phenomenon could operate on the phenomenon through the thing itself, which is in all phenomena one and the same thing" (Schopenhauer, 1995: 20).

To this extent, it is accepted that with the language one can do things, but one can also operate with power over the things. The magic operates changes in the world following a different way to the one of cause-effect. In this way a fundamental means would be the language. In this direction are significative the findings of physics and cybernetics. "Life is a system of conduction and analysis of signal that allows the recognition of the high life, the one, God within the narrow womb of the evolution of matter. It is a system for receiving God's image and its amplification in the universe of matter", (Carvajal, 1988: 33).

The human body needs primarily energy to do things. This energy will come through packets. But in the cosmos, energy has also a way of moving, that is, in waves. These can be mechanical or electromagnetic, without excluding other types of waves, not yet registered by the science. Within the mechanical waves are the ones of sound. Spoken language contains such frequency of wave. It can transmit a certain energy to the things toward which are addressed, until operating upon them some change. In that sense, one could speak of a force of enunciation corresponding to the act of enunciation of operating. Over it, the magical action would be based.

The emission of the articulate voice includes the following three effects of simultaneous action:

1. The emission of a sound puts into action the material plane of nature.

2. The emission of a certain portion of vital force puts into action the astral plane.

3. The release and creating a psychological entity, it is the idea to which the sound gives a body and the

articulation to life. Every idea created of such way, and so being manifested in the material world, acts for a time as a positive being. Then, it is going to extinct and disappears gradually, at least on the physical plane. The duration of action of this idea depends on the cerebral voltage which it has been emitted. That is, the sum of vitality with which is provided (Papus, 1988: 169).

From the characterization of the healing acts, all magical acts have their common denominator in acts of enunciation of operating. The semantic rules for the use of any indicative device of force of enunciation to **operate** OP (**speech acts to operate**) are:

Rule 1. OP must be issued only in the context of a **magical text** MT, whose emission predicates some **magical power** MP of magician M. This would be a rule of propositional content. It is derived from the conditions of propositional content number two and three.

Rule 2. OP must be emitted only if the user U prefers that M applies MP to that not to apply MP, and M believes that U prefers that M applies MP in front of not to apply MP.

Rule 3. OP must only be emitted if it is obvious to both M and U that M will not apply MP in the normal course of events. The rules number two and three are preparatory rules. They derived from preparatory conditions number four and five.

Rule 4. OP must only be emitted if M intends to apply MP. This is the rule of sincerity. It is derived from the condition of sincerity number six.

Rule 5. The emission of OP accounts as the assumption of an obligation to apply MP. This is the essential rule.

With these semantic rules one can analyze all types of magical acts. In the orison-magical-action, for example, the preparatory conditions include:

1. The magician is in possession of a power

2. The user asks help for an infirmity of physical origin

3. The sincerity condition is that the magician wishes to remedy the cause of physical pain

4. The essential condition is that the magician propitiates with the whisper of his recitation of the magical text, that is, communicates to the user certain energy capable to operate immediately on the origin of physical pain.

The herbal-magical-action has as preparatory conditions:

1. The existence of a person, eager to subdue another through the services of a magician.

2. The sincerity condition is met, the magician accepts do it.

3. The essential condition is given by the emission of some magical texts. In them some energy is transferred to some objects. These will produce distortion in the conduct and analysis of information for the person to whom it is addressed.

The action contrary, that is, the de-herbal-magical-action, would have the follows preparatory conditions:

1. The existence of a person with distortion in the conduct and analysis of the information required for the adjusted functioning of his body. He goes for help where the magician. He knows how to enter a rearrangement of information.

2. The condition of sincerity: the magician wants to make such a rearrangement of information.

3. The essential condition: the magician seeking with the emission of some magical texts to transfer energy. It aims to eliminate the distortion factor set by another magician in a few objects for altering the conduct and analysis of information in the body of the user.

In this regard, it should be understood the initial hypothesis. Magic is objectified language with the imagination to act with power. The magic in its experimental period is trying to clarify how this happens. Everything points to show the language as privileged means to transfer that energy. Beyond a conceptual network called magic, a scenario not explored by linguistic is opened. This shows a quality of language so far ignored: **the one of operating over things**. It comes to join those already studied like the one of designating and doing things through of saying or with words.

This quality of language has already begun to be explored intuitive and practical way because of the magic. They are worth noting in this field the work done by Bandler and Grinder in the United States. These researchers, from the generative-transformational model by Chomsky, have found a way of doing therapy following the grammar of the language used by patients. Not surprisingly, their texts were titled with name: Magic I and II, where they publish their results and procedures.

"We are proposing the existence of a subset of sentences in English. We recognize them as well formed in therapy, and they are acceptable for us as therapists. This set of sentences satisfies the following requirements: 1) they are well formed in English. 2) They do not contain transformations for elimination or eliminations by unexplored in the part of the model where the patient feels he has no alternative. 3) They do not contain nominalizations (process-event) [events taken as incidents occurring at one certain time and then end]. 4) They do not contain arguments words devoid of benchmark indexes. 5) They do not contain verbs incompletely defined. 6) They do not contain unexplored presuppositions on the part of the model in which the patient feels he has no alternatives. 7) They do not contain phrases that violate the conditions of well semantic formation", (Bandler and Grinder, 1980: 79).

Alterations are identified in the conduct and analysis of information by alterations in language. These are expressed in use of language. By rearranging expression in the use of language, a rearrangement is achieved in the information. The patient is oriented to produce adequately the sentence altered in its grammar, and evil disappears immediately. The organism needs of it for a rearrangement of system of the conduction and analysis of language. This conception of language goes beyond the position of doing things with words. It is to operate over things and over the man through

131

language. Initial studies, of Bandler and Grinder, after led to the formulation of Neuro-Linguistic Programming (NLP)

Now, if this analysis done on the magical acts have a certain foundation, according to the occurrence of phenomena, this would imply additional formulations for language theory and the theory of action. It is necessary to cover these fields of experience. Man's knowledge will be greatly enriched.

With all these considerations, it is easy to understand the magic as a system. According to the block diagram representation would be as it is shown in the image 4 (see this picture at the end of this book, for a better understanding of the explanation).

The system is activated by the recognition of an initial state as a fragment of world history where confluence the seeking by help from a person, the expression of corresponding symptoms of a sorcery and the receiving by a magician of the expression itself, under certain social and cultural conditions, which also has the experience of each of the participants in this initial communicative event, consisting of a knowledge and a system of prior beliefs, and a system of conventions between magician and user.

With the expression of the symptoms of the sorcery, the magician can build a semiotics of the sorcery. He takes also elements of the context in which the communicative encounter is given, and the experience of the participants. With this, the magician comes to making some decisions. The first semiotics is subjected to a data processing determined by the capabilities and powers of the magician according to his magical knowledge and training in the magical activities. This allows him to perform a translation of the initial symptoms. He effects a change in semiotics where these symptoms take on another meaning. They take another way to be coded to facilitate reception according to certain cultural parameters.

With these operations, the magician is in possession of a data packet. This leads to choose procedures for conducting of the same. During this data processing, the magician stored in his memory the decisions initially taken and the procedures for reuse them in case of corrections, according to the results. This choice of procedures leads to the execution of specific magical action. It corresponds to a solution for the semiotics of the translated sorcery, and therefore, a second communicative event.

It may be the case, this single magical action is not sufficient to achieve relief from the sorcery suffered by the user. In these cases, it is needed a feed back into the system with the results, plus a collection of external data to take a new decision. Reprocessing of data is achieved, that is, a third communicative event. An example, about this, is in the first slug dumped to toilet. This forced a new query to Don Pedro in front of the relapse of the patient. With the new data, he does a feed back into the system. He obtained as a result that he needs a semiotic more precise or eliminates noise in the emission of procedures. In this case, it was the elimination of noise in the emission of procedures, that is, slugs should be burned and not dumped to toilet. Is only logical, before he reaches this result, he analyzed the decisions and the procedures followed previously. He looked for possible errors, and not finding them, he explored in the environment to find the cause of dissatisfaction with the results.

It has been thus presented in the light of magic as a system, the different steps followed in the same:

1. The expression of symptoms

2. The semiotic translation

3. The execution of the magical action to obtain results

4. Feedback into system.

The image number 4 summarizes in some detail the actions of magic.

PART FIVE: REGISTRATION OF MAGICAL SPEECH

Luis Carlos Molina Acevedo

The Magical Texts

Now, all interest will be directed to the study of magical texts to establish, from them, the structure of magical discourse, according to the methodology proposed by Teun A. Van Dijk (1981) for discourse analysis.

Fixed inscriptions of magical language are in ordinary language named **magical formulas**, **magical orison**, **magical secret**, and **magical recipe**. Each of these designations correspond a different configuration of language. They can be studied from a general theory of discourse. The **magical orison** is the main linguistic instrument for the **prayer-magician**; the **magical formula** is the one for the **helper-magician**; the **magical secret** is the one for the **healer-magician**; and **magical recipe** is the one for the **herbalist-magician**. There are some other names from ordinary language for magical texts. They correspond or are located in one of the four previously defined categories such as **exorcism, charms, spells** and **incantations** used by the helper-magician, and **prayers** and **petitions** used by the **prayer-magician**.

Superficially, the four categories of fixed texts of magical discourse, they differ because:

1. The magical orison, as such, stands at the mere recitation of the text.

2. The magical formula acts as an invocation of supernatural forces. They are of diabolical origin. They require additional rituals.

3. The magical secret mixes the orison with the action of some active ingredients from animals, plants or minerals. It describes how to proceed with such principles. These act of magical manner in view of the casual observer.

4. The magical recipe gets its power from the mixture of natural elements, whose magical procedure focuses on the collection of components under certain conditions of time and space; they are applied to the preparation of the magical recipe.

The Magical Orison

As an example of this kind of magical discourse can be cited the following:

SECRET TO GET STAGNANT BLOOD

"When our Lord Jesus Christ passed through Rome, the holy house of Jerusalem; saying, stop, stop, stop blood", it is said three times and pray a creed putting three leaves on the wound and outstretched hand; if it is nasal, it puts the hand on the forehead and if it is vaginal, it puts over abdomen and a creed is prayed and if they are several wounds, it puts a white sheet and hand extended over wounds and it is said three times and a creed is prayed (Anaya Campbell: 170).

You can be distinguished in this text various component parts:

1. A **title**, which acts as a discursive macro-structure, that is, such as a summary of the content of the text, but it acts also as an identifier thereof.

2. A **recitation**, which is the text actually recited. It is marked in quotes. It encloses the power of language.

3. A **procedure**, which is dedicated to the description of the variants for recitation and complementary actions. It tells how and under what circumstances to operate in a certain way. It indicates also which common prayers should be added.

The first part is marked by a linear array of information characteristic of all such texts. It is emerging as a discursive macro-structure and as a bounding over other texts. It should also be noted, the information management is made based on the action. The title is a global discursive macro-structure of the text. It indicates what is done with it. Here, the designation of "secret" does not refer to the structure of the text but to the condition of use. It indicates the characteristic of being a text to act with power over the world.

The second part consists of text to recite. It contains all what should be said to act with power over the circumstances of bleeding. This element of the structure of the text is characterized by the evocation of an effective historical-religious event. The helper-magician intends, by analogy, to replay it as an actual action. He tries that again happens the same in a now-here, by the mere fact of reciting what happened in the past time. The biblical origin of this orison is obvious. Therefore, it is difficult to pinpoint the historical moment of creation of this text. The origin of such a discourse made at a fixed text of magical character, is lost in time. The recitation structure consists of a phrase plus a clause. They represent the presupposition, the known. It is connected, by a semicolon, with another phrase. It fulfils the function of the introduced, that is, the new in terms of information.

The clause embedded between phrases, it has as function the one of clarifying that Jesus did not go through Rome. He goes through Jerusalem where Rome had its representation. It was the centre of the Christian religion. On the other hand, while the first period of the sequence of sentences refers to the description of an action, the second period of the sequence is an illocutionary act of ordering. The second sentence uses positive discourse of illocutionary force of ordering within the new information. Such a force extends to the entire sequence of what is said. At this point, it is where would play a great role the definition of a third function about language, which is understood as a capacity of operating over things, next to the functions of designating and of doing things with words. To that extent, it would be understandable, in certain linguistic practices of man that orders are given to things, that is, to the blood in this case, which lack of will to obey. Without this feature of language, magician presupposes that magic has no reason to be.

The third part of the structure of the text is devoted to instructions. They should be followed during the recitation of this. These instructions are of several types:

1. It goes to the recitation itself, "it is said three times"

2. Then, other instruction focuses on the complementary recitation, "and a creed is prayed"

3. A third type of instruction specifies under what circumstances and conditions should be applied the recitation.

With these elements, one can proceed to make abstraction to apply the categories of text called magical-orison: **title**, **recitation** and **procedure**. The title corresponds to an overall macro-structure of the text. Recitation corresponds to the evocation or drawing a parallel with a historical religious event. It is intended to update it with the action of a ritual. And the procedure indicates the instructions for performing the ritual and expressing the recitation, which allows identify the follows subcategories of procedure:

a. Complementation.

b. Expression of what is said.

c. Condition.

It should be clarified that these categories correspond to the text of the magical-orison. The prayer-magician uses this in his practice of social function. It is observable how ordinary people pray also some orisons of these, but removing structural elements, which can only restore by magician. The most elided parts are the instructions.

As a further example of such texts may be cited the follows:

SECRET FOR DISLOCATION

Halt ferocious animal, bow your beard on land; before you were born on earth was Mary. I am the rod of Reon, under the Holy Spirit.

Saints come from heaven and the Church come saints and the holy land come also saints. It is said nine times. For the animal nineteen times (Anaya Campbell: 169).

It is here also noted a title. It presents an array of information depending on the action or magical act. It is a global macro-structure of the text. Then, it puts the recitation with a clear allusion to the Bible where the devil is seen as the cause of the dislocation, and therefore, as something that can be dominated with a higher power like the one of the Virgin Mary. Up to this point, it is presented a preparation of the fact by a sequence of phrases. Then, it puts the historical-religious fact. It is updated with the ritual as an essential condition. Then, in the second paragraph, the sincerity condition of magical action is identified. It is a propositional content not explicit, restorable well: not only from the sky and church come saints, they are also possible the existence of earthly saints or magicians, If I am a magician, then I can do that occurs what was done with the rod of Reon under the omnipotence of the Holy Spirit.

Third, it is the procedure. In this case, it is reduced to the subcategories of expression of what is said and of condition. It has been deleted the supplemental instruction. Anyway, the prayer-magician will recover it at the macro-structural level. At least, that is observed in practice from prayer-magician. He always accompanies the orison or recitation with different Hail Mary and Lord's Prayer, while he draws crosses over the dislocation, with his fingers.

For this kind of magical text, one can find titles like existing evils, for example: "For bad urine",

"Against the gout", "Against burns," "Against evil breasts", "Against the cataracts", "Against the evil of Saint Pablo (epilepsy)", "to cure worms", "for nightmares", "for the bad times", "Prayer to upset women", "Prayer to help expel the placenta", "Prayer against tapeworm" and "Prayer against cancer", among others.

The Magical Secret

It has seen said before, this activity corresponds primarily to healer-magicians. It was noted how the main function, of it, it is to cure the bites of poisonous animals, essentially snake bites. This type of text is presented as the most complex because of the multiple structural elements included therein.

In a magical-secret to cure snake bitten, a combination of empirical knowledge and magical knowledge is observed. For this analysis, I will consider the expressed in the book "Secrets and Recipes" prepared by Eduardo Anaya Campbell. It is a discourse too detailed. It is similar, in structure, to those by Uribe Escobar (1967: 45-52), another major compilers of such texts in the culture of Antioquia. The magical discourse, about the treatment of the snake bite, begins with an introduction where a story about the treatment of this disease is made. Then, they are identified the kinds of snakes known in the world, and the most poisonous insects like centipedes, scorpions, poisonous spiders and stingray.

It is also necessary to point out which are snakes and insects existing in Colombia, and snakes, which are poisonous and which are not. The same is done for departments such as Bolívar, Boyacá and Cauca. Then, it proceeds to present the different methods used by healer-magicians to cure snakebites (p. 21-35). After that, a list of the **cons** (remedies against snake bites) most popular in the country is included, where the popular names of plants are presented. They act against the venom of snakes (p. 35). Then, the formula to prepare the gall of snakes, or serums made from each poison of animal is introduced (p. 35-38).

Complementing the above, there is a preparation of tinctures. They come from bitter plants as an active ingredient to cut poisons (p. 39). Then, he notes the semiotics of the symptoms present in the patient. He specifies what procedure to follow in different cases. It is also taught to read the degree of poisoning from the symptoms, what kind of snake bite occurred, and what is its healing (p. 39-44).

The introduction is followed by what is called "The map of Salvador Elias". It refers to the practices and procedures of this healer-magician in particular. The specific procedure for the bite of each snake has also considered plus the status of the snake: if it was pregnant at the time of biting, or neutered, or refined (with skin freshly molted), or fasted (hungry) (p. 45-58). But, the remedies should be applied according to the size of the snake (p. 58).

It should be clarified that all this is restored through the observation of the patient. Most of the time, he may be unconscious. He can not give reliable information. It should be released from the aspect of the bite and the patient's condition. The healer-magician should be a highly trained person, by the proximity of impending death, resulting from the bite of a poisonous snake. Then, they put the charts of remedies, which offer alternatives for the cure. This is important for the adaptability of the method. Not all plants grow in all climates. Here, they are again important correlation tables, no longer in a hermetic sense, but in a practical sense. Knowing where to find the active substance, independent of place and time, is essential for the universal magical practice.

The healer-magician must follow the procedure, according to the semiotic reading done from all elements (p. 59-62). Then, he talks about the food for the patient with snakebite (p. 62). There are also procedures about how to cure complications and post-healing effects: cramps, vomiting, sores and other (p 62-76.). It is also presented the mode of preparing compresses (p. 77-83). These are applied to extract bites poison.

This detailed presentation of folk medicine to treat snake bites, ends with the following paragraph:

"A bitten person is cured only have faith and hope, always looking for the natural roots of cons without distrust any of them nor of the will of God, or loving Mary and the Anima Sola (single soul) and fear God, who is the one can do all and do upset when he wants; He gives power to those who want and love the art of S.S. Pablo, the teacher of curing snake bite" (Anaya Campbell: 80-81).

Incredibly, after such detailed exposition, enough to cure any snake bite by the wealth of empirical knowledge shown by the scholar, he culminates with a magical accent, almost denying all what was said before. This factor becomes to a heal-magic-action in a magical practice and not as a practice of non-traditional medicine as it should be.

By the one exposed about the magical secret to cure snake bite, it is clear that this aspect would only look for extensive linguistic research about the theme. Therefore, in order not to prolong this analysis too, it will be presented, going to a higher level of abstraction, the structure of the text called magical secret, as follows:

1. A generic title, which globalizes in a macro-structure the content of the text.

2. An introduction where the characteristics of the healing art of snake bite are set.

3. A procedure where knowledge of the snakes and the semiotics of symptoms occurs.

4. A treatment, which is an inventory of the procedures to be followed after the translation of the symptoms to a healing semiotics.

5. A chart of remedies where recipes are listed themselves, to pursue with additional indications.

6. Magical reinforcement, where reference is made to the supernatural forces to heal the bite of poisonous snake.

Now, the categories set forth above may be fused with each other for access to an even higher level of abstraction and to classify categories of healing discourse applicable to all magical discourse of this kind. It has accordingly, title, introduction and explanation can be merged in **explanation**; and treatment, chart of remedies and additional indications can be merged into **treatment**. In this way, it has a structure for curative discourse comprised of:

1. Explanation

2. Treatment

3. Reinforcement

The structure of healing discourse provides sufficient elements to establish the difference between healer-magician and prayer-magician. While the healer-magician will give equal importance to all three elements of the discursive structure, prayer-magician focuses only on element of reinforcement. Thus, the categories established for the gender of magical-orison become subcategories for the genre of magical-secret. The prayer-magician facing the same snake bite will only do the following:

SECRET TO HEAL

For San Pablo and San Benito was conceived this orison and on behalf of these saints I do this revelation.

Venom and poison die; venom and poison heal inside and outside die; outside and inside, they heal.

This orison is repeated three times and each time a Lord's Player and Hail Mary Gloria Patri should be prayed.

Note: When it is to cure a snake bite or other poisonous animal, it is repeated three times making the sign of the cross over the bite and praying the orison, then the Lord's Prayer and Hail Mary (Anaya Campbell: 88).

Instead, the healer-magician, despite of saying also the orison, does also the other steps. The healer-magician, besides knowing the language settings, which will operate over the evils of the user, he has also knowledge about the active ingredients from natural elements, little known to ordinary people. To that extent, this type of texts is called magical-secret. It contains secret knowledge in two directions (linguistic and herbal) and therefore it is doubly magic.

The Magical Recipes

The magical-recipe as magical discourse shares certain features with the magical-secret. It takes the combination of certain elements. But unlike the heal-magical-action, the cure is given by the active ingredients from the components themselves. In the magical-recipe, natural elements receive the power to operate by the magician. They are added qualities through the magical ritual. In the herbal-magical-action, it is applied a mixture of the elements to the user. This is accompanied with orisons. That is the essence of the ritual of the heal-magical-action. The herbalist-magician, however, applies the ritual to the combination of the elements, not to user. The orisons are applied to the elements, not the patient, to give them power. As an example of a magical-recipe, it can cite the following:

A POWERFUL TYING BY HAIRS

A small bottle of amber colour –it is found in pharmacies

Hair of the beloved

Some saliva from you

Get hairs of the beloved -you know how to manage to get them- on new moon, you enter hairs into the bottle and add a little of his own saliva. You close the container, keep it inside your pillow, you make a cover and into it you place the container, and you sew the pillow again. With this spell it is achieved the all domain of the person of your interest, as is known hairs represent a strong bond of energy to get what you want in a short time. It added that when his zodiac sign belong to Taurus, Virgo or Capricorn, the bottle should be buried closed- -within the container in a pot; in a public garden or in your home, if you have one. This is done due to the affinity of the above signs with earth, which is a basic element for the successful completion of the work (Lara, 1989: 20).

Here, a **title** is noted at the structural level. It acts as a global macro-structure of the text. The information is sorted according to the magical action to be performed. It is highlighted, in this sort of information, the initial modal. That "powerful" denotes the existence of different magical-recipes to tie with hairs, but none is as powerful as this. In turn, the linear macrostructure of title has a deletion in the information. A semantic projection of it would be: powerful system to tie the beloved by using hairs from him.

Secondly, it is observed, at structural level, the enumerating of **ingredients**. It is noted, the components are not appreciated by their active ingredients. Here, the interest is in the powers added with the magical ritual, with the preparation of magical-recipe.

Third, it is also included the structural element corresponding to the category of **procedure**. This in turn has sub-categories as follows:

1. Procedure for getting the ingredients

2. Procedure for mixing the ingredients

3. Procedure for applying the magical-recipe

In the subcategory of combination, it should be noted the presence of "new moon" element. This marks an atmospheric condition for the magical power of that component and the scope of its effectiveness. The same should be seen in the subcategory of application, where the importance of the zodiacal signs and their affinity with the earth is highlighted. This adds more magical power to the compound.

However, it should be noted an arrangement of specific information. It is embedded within the category of procedure in which reads: "With this spell it is achieved the all domain of the person of your interest, as is known hairs represent a strong bond of energy to get what you want in a short time." This corresponds to a preparatory condition. It is a preparatory action for the implementation of essential act. It is something like a justification by inference about why is important this magical-recipe.

There is, then, the magical-recipe as a magical text presents a structure of text-type corresponding to the following categories: **title**, **ingredients** and **procedure**. In turn, these categories become subcategories of heal-magical-action. These are included in the category treatment. But, it is also included the category procedure into the magical discourse called magical-recipe, which, then, has three sub-categories: procedure for the obtainment of ingredients, procedure for the combination and procedure for the application of it.

It should be noted that these are some structures of modelling for the study of magical texts. But it is possible that in the inscription of texts not appear all categories mentioned here for different types of magical discourse. However, they are present in the practice of the various magical acts. This can be explained by the need of the magical discourse of being hermetic. To maintain certain ethical principle in the exercise of magical craft some textual items are elided. Therefore, although the inscription of magical discourse constitutes to it in a scientific object, such a study should be complemented by empirical evidence, if one wants to find those missing elements in the fixed texts. By way of further example, one can cite the following magical-recipe:

RECIPE TO CURE SYPHILIS, PUSTULES AND SORES OF BAD CHARACTER

You take two ounces of potassium iodide and they are given in one litre of water, taking three daily cups; at fifteen days, it is suspended for three days, the patient bathes and takes a purgative composed by 2 ounces of England salt and retakes 2 ounces more of potassium iodide, in the same way above indicated and turns to bathe and take another purgative as before. The patient will appear to put very ill because him weakens and sustains bad taste in the mouth and above all, he burst in sores but by that the remedy will not suspend because the sores heal all at its time.

Then, it takes a pound of bush of vine, it will crack the long and does 9 bunches and it is boiled one every day early in 3 wells of water until they are reduced to one, and that is the breakfast for 9 days and after are re-boiled throwing 2 for 1 and drinks; after that, he bathes and will purge as stated above. The patient will already be good in sores, but again repeat all remedies even if the patient appears healthy until the scars of sores that are purple, turning once wear white and already the ill will fatten and taking off the scars from the face and on the body and the whites of the eyes will be like of the children, that is, bluish. (Diet: no fat, no fermented beverages; no fish or pork rinds) (Uribe Escobar, 1967: 52-53).

In this example, one can see more evidence the character of macro-structure given to the titles of the magical texts. It is noted how this **title** presents an ordering of information based on the magical action. The text has really three magical-recipes in one. It is reflected from the same title. This magical-recipe corresponds to the practice of the healer-magician. It has the same structure that one from herbalist-magician. In this, it is only missing the subcategory of procedure for the obtainment of ingredients. Otherwise, it can be said, categories: title, ingredients and procedure are subcategories of the category of **treatment** from curative text or magical-secret. This shows also that the only structural difference, between magical-secret and magical-recipe, is the subcategory: the procedure for the obtainment of ingredients. It refers specifically to conditions of time and space. These added magical properties to components.

Luis Carlos Molina Acevedo

The Magical Formula

This type of magical text marks the scope of helper-magician. It will be the subject of a study more extended because of its scope of applying. The magical-formula is dependent on the power of language. It takes certain symbols to impart a special power to ritual. As an example of the analysis, it will be taken the sortilege: "Orison to Saint Helena" in different versions. The reason for this is due to the widespread use of it by the inhabitants of Antioquia and the multiple uses given to it.

In Chapter III of the part called "The Black Hen" from "The Book of San Cyprian" reads: "The sortilege that is formed through practical and magical ceremonies, using them to achieve those things which by natural means would not be us possible" (p. 111).

SORTILEGE TO TIE A PERSON

For this sortilege is needed to prepare a medal of St. Helena, placing it on a rag green silk, in which three small gold nails are nailed, which will be used for the ceremony. An object is also needed, portrait or figure, which is dedicated to the person who wants to tie, in which it is to be nailed one nail as indicated in the following orison:

O1

INVOCATION TO SANTA ELENA

O glorious Santa Elena, loving mother of great Constantine, the Roman emperor! Thou, that being the daughter of the king and queen, went to Mount Olivet by yours endearing love toward the divine Jesus.

I require yours powerful intercession to get what I want. Of these three nails of Our Lord Jesus Christ, in imitation of those that you had, one I give it to your son, the Great Constantine alike is in your holy image, another I throw into the water as thou throw it to the sea, for the salvation of sailors, and the other, I nail to this object dedicated to N (name of victim), to be nailed in his heart, so he can not eat, nor in bed to sleep, nor in it sit, nor with woman or man to speak nor having time to rest, until that, by thine intercession, he comes surrendered at my feet.

If this I wish it was granted me by yours mediation, all my life I will be yours sincere and devoted lover, for ever and ever. Amen (p. 111).

It is all too obvious the structure of the magical-formula. Maintaining parallelism in the analysis of other magical texts, and attending to other versions of this sortilege, the following categories are proposed for this structure:

1. Title

2. Recitation

3. Ceremony

Regarding the magical-orison, it differs for filing a ceremony rather than procedure. This ceremony includes symbols.

In this sense, the **title** no longer is an ordering of information according to the action, but in terms of the supernatural force invoked. The helper-magician needs allied spirits to carry out his works, in this case Santa Elena. Here, it is surprising to note the case of Dona Rosa, mentioned above. She in his practices of black magic, not only invoked allied spirits through orisons. Besides, she gets friend spirits in cemeteries. She has at least five in each of the cemeteries of Medellin. They help her in her works. She is permanently renewing the staff of spirits of the dead. She says that they can not be spirits of people killed. It serves only those who died by natural death. She goes to funerals. She asks among the attendees how he died. When she finds a candidate, she waits until when people leave. Then, she gives a knock on the grave and asks John Doe, can you help me? He will answer yes or no. In the affirmative case, she recorded the name. Then, in her ceremonies at midnight on Tuesdays and Fridays, she will call them at eleven o'clock at night. She receives them with liquor, cigarettes and a radio tuned to a musical station. At twelve o'clock, she will start the work. Before that, she will talk to them about characteristics of job.

As it was said about the category **recitation** of the sortilege, one can identify the following subcategories:

1. Identification

2. Actualization

3. Farewell

In the subcategory identification, they are the most salient features of the spirit invoked. The actualization gives the reference to the historical-religious fact. The helper-magician intends to replicate this fact by executing the same steps. He hopes to obtain the objective pursued by means of the spirit invoked. And the subcategory farewell, is a kind of thank in advance to the spirit. In this case, the explanation is facilitated. Each of the subcategories is corresponded respectively to one of the three paragraphs, components of the orison.

The ceremony, in turn, presents the following subcategories:

1. Preparation

2. Disposition

3. Execution

Preparation is the obtainment of the indispensable elements for the ceremony such as nails, the medal of St. Helena, the green silk and portraiture. The disposition reflects the physical and mental conditions required by the performer for the operation. Here, they are presupposed by having been treated in another section of that book, that is, in the "Book of San Cyprian". This section is called "Essential qualities to profess the magical arts", numbered as Chapter V. Anyway, it will be quite illustrative the voice of Dona Rosa who says: you can not eat meat during the three days prior to the execution of the ceremony. That day can not eat beans after six at night to achieve greater efficiency.

Subcategory execution refers properly to steps to operate in the ceremony. Here, it should be clarified that the sortilege studied is part of a work of magic. Therefore, it must be assumed, at this point, that the magician already knows other aspects required for the ritual.

O2

ORISON TO SANTA ELENA

(Sortilege to tie a person)

Take a medal of St. Helena and you hang it on a green silk cloth, in which three small gold nails are nailed. Then, you take up an object, figure or portrait of the person you want to tie, in which one of the nails is nailed under the following invocation:

O glorious St. Helena, mother of Great Constantine, Roman emperor! Thou, that being the daughter of the king and queen, went to Mount Olivet by yours endearing love toward the divine Jesus. I require yours powerful intercession to get what I want. These three nails of our Lord Jesus Christ, in imitation of those that you had, I dispose of them in the way that thou did it. One I give it to yours son Constantine Emperor, so it is in yours holy image; I throw another to water as thou throw it to the sea for the salvation of the sailors, and the other nail in this object dedicated to ... for what nails in his heart so that him can not eat, nor in bed to sleep; nor in chair sit, nor to speak with man or woman, nor having time to rest, until that, by thine intercession, he comes surrendered at my feet. If this I wish it were granted me by yours mediation, I will be all my life yours most loving and sincere devotee, for ever and ever, Amen (Anaya Campbell: 162-163).

Although in this version, they are not major changes in relation to the content of O1, it should be noted a relief of different information. First, titles of O1 are merged into one in O2, so the first title becomes a subtitle. It keeps the denomination most important for the enunciation of this second magical text, according to the spirit invoked. Another change in the ordering of information is given in the category ceremony. While O1 uses the description, O2 uses the narrative. This indicates a more popular character of O2.

Regarding to the actual invocation, here referred as category of **recitation**, they should also be observed differences. While in O1 each of the subcategories corresponds to a paragraph, in O2 all information is equally important. It is presented in a single paragraph. That is in line with the popular character of the enunciation of O2 noticed before. The orality has fewer tendencies to make the separations made in writing. To this point, one can say that O1 is the source of O2.

O3

ORISON TO SANTA HELENA
(Sortilege to tie a person)

Take a medal of St. Helena and you put it on a green cloth, in which three small gold nails are nailed. Then, you take up an object or picture of the person you want to tie, in which one of the nails will be nailed in according to the next invocation.

Oh, glorious St. Helena, mother of the great Constantine, the Roman emperor. Thou, that being the daughter of the King and Queen, went to the Mount of Olives by yours endearing love toward the divine Jesus. I require yours powerful intervention to get what I want. These three nails of our Lord Jesus Christ, in imitation of those that you had; one I give it to your blessed image another I throw to water like you throw it, to the sea for the salvation of sailors, and another nail in this object dedicated to N.N. for being nailed in his heart, so he can not eat, nor to sleep, nor in chair to sit, nor with woman or man to speak, nor having time to rest, until that by thine meditation, I will be all my life yours most loving and sincere devotee. Amen (Kardine, 1975: 72).

It is evident, in this inscription of the text, despite retaining the same structure of O2, how the first content changes appear. In O3 is no longer emphasized the term invocation. Regarding O1 and O2, the father of Santa is not an indeterminate king, but a king known to the speaker. He does not need the intercession but the intervention of the Holy woman. He demands a more direct help. They have also disappeared marks of exclamation at the beginning of the invocation. This corresponds to the absence of emphasis with the term invocation. O2, relative to O1 and O3, has the addition of the phrase: "I dispose of them in the way that thou you did it." However, in O3 there was deletion of the phrase: "until that he comes surrendered at my feet". In addition, it is changed intercession by meditation, in a clear correspondence with the demand for more direct intervention by the Holy woman for greater efficiency.

O4

ORISON TO SANTA HELENA

St. Helena went to sea,

Cross and nails went looking;

Make mends me, St. Helena,

What we will find.

O glorious St.! You are mother and queen of the Passion of Christ. Roosters already crow, and the earth trembles, and the sun sets.

The day already darkens, and the temple veil is torn, The Savoir already dies on the cross, and the Virgin sees her Son die. How painful! And the glorious St. Helena sympathizes with the Virgin and comforts her and says, "I'll help in her pain and continue to Mount Calvary in search of the three nails, the three powers, the crown of thorns and the cross of the Redeemer". Everything found at the foot of the cross; she knelt and wept, the five wounds washed and kissed; the three nails disengaged; the cross caught with her hand for the conversion of those who are carrying the cross on their back and carry it should not take.

The three nails divided, leaving one for her; the first throw into the sea by the conversion of sailors; the second gave it to the Emperor Constantine, her beloved husband; the third lent it to Our Lord to take it on the cross for miracles to be asked by his miraculous nail.

(What does the glorious St. Helena? Bring up what is lost, makes paying bills, gives health to the sick, brings men and women of work in which they are, does come to that is absent from house, whoever it may went; this orison is for anyone that we need, so this person do not like or pleasure, nor to sleep, nor to eat, nor to drink until this person comes where the person who does the request).

I pray thee, glorious St. Helena that by thy great power helps me to get through Our Lord that I smite with thy powerful protection in all jobs. I ask thee this through Our Lord Jesus Christ who died on the cross. Amen.

Secret of the Orison

To achieve the grace of a patent miracle from her, it must make three sacrifices: get three candle ends as follows: the first is asked to a homeless person, and placed on the left; the second, it is asked to the church, and three nails will be shaped in cross and placed in the centre; the third, it is purchased from a greedy person and placed on the right, and so will be presented on Good Friday, the day of the death of Jesus on Calvary on the cross between two thieves. The time when orison is made is that of the twelve; knowing it of memory can be done at any time, always with the same ends; but worth should be on his knees with open arms at the foot of the statue of Saint Helena of the blessed cross, one person, closed, ending with one Our Father and one Hail Mary to San Alberto, one salve to the glorious St. Helena and one Our Father and one Hail Mary for the relief and rest of the souls in purgatory. Amen (Uribe Escobar, 1967: 111).

In addition to this text, it is interesting the footer note, placed by the compiler:

Orison to Saint Helena, this pylon of nonsense and historical falsehoods is supposedly the orison to Santa Elena, which we decline to comment as to its grammatical wording is stupid. The downside of this is not so, but that circulates in printed flyers to the price of five nickels, so that humble people make use of the grotesque that is neither orison, nor anything. As for the secret of it, as well as by the thread is known the ball, so by this secret one can see how far are mixed chicanery, fanaticism and the notorious art. The Lord's Prayer that is recommended to San Alberto Magnus, qualified as sorcerer for all demonologists, and as for its accuracy historical-chronological, better leaves well alone. This orison has foot of imprint by the Catholic Typography of this city (?) (Uribe Escobar, 1967: 111).

It shows in this inscription of the orison, the presence marked of the collective. Just look at his recitative tone to check that is directed to be memorized. The verses at beginning well demonstrate it, besides being expressed by the stile. The rest of the text follows the guidelines of popular poetry in terms of loudness. It seems that memory is more receptive to certain repeated resonances, like the rhymes. Anyway, one should look somehow the text retains the essential content and intention to obtain an end. The least important for people is the historical veracity of an orison. It matters is its effectiveness.

Neither version of the orison to Saint Helena fits to the historical reality of the facts, but the compiler defend the authenticity of some. Here, it is where plays a role the life of Santos. In this, it is most probable that is inspired this orison. Do not forget that was the Church who has spread the legend. For this, St. Helena found the cross on which Jesus was crucified. Among the many conjectures about it, it is worth quoting the following, for example:

St. Ambrose and St. John Chrysostom inform us, excavations began at the initiative of Santa Elena. They resulted in the discovery of three crosses. The same authors add, the Cross of the Lord, was among the other two. It was identified thanks to sign on it. Moreover, Rufino, who follows to Socrates says, Santa Elena ordered excavations should do in a given site by divine inspiration. There, three crosses and an inscription were found. As, it was impossible to know to which of the crosses belonged the inscription, Macarius, Bishop of Jerusalem, ordered, be brought into the site of the discovery of a dying woman. The woman touched the three crosses. She was cured in contact with the third. So, it could identify the Cross of The Saviour (Butler, 1965: 205).

This orison O4 has been modified by the oral tradition. It is evident not only in tone, but in the pedagogical way as worded. In this sense it presents a high functionality embedded parenthesis in the text, where multiple uses for this ritual or ceremonial is listed. Still, there are uses for mentioning, such as move away to unwanted neighbours. It is also used to find lost things with lights only a candle end. When it is consumed, with the candlewick indicates the direction in which is the thing. If it points the exit of the house, it will be interpreted as that the thing was stolen. All these features have made of this orison and its ceremonial, a widespread practice among people, even among very Catholics.

O5

ANOTHER ORISON TO THE SAME HOLY WOMAN

Powerful Saint Helena! Daughter thou are of King, daughter of Emperor; daughter of the Great Lord; the cross thou searched and three nails thou found. One of them thou throw to sea, that was with which thou find it; another thou gave it to your son Constantine, to be rid of war and battles, and the other thou have in your possession. I do not ask thee to give it to me, but to lend it to me to be nailed into the heart (hence the name of the person you want) to arrive at my feet, meek, three times (Escobar Uribe, 1967: 113).

The compiler adds a footnote: "This, at the same St., is from the collection of Dr. Caballero Sierra; has some more good judgment and, as above, it belongs to the notorious art. Note that in the above, Constantine is 'her beloved husband' and in this his son" (Uribe Escobar, 1967: 113) is.

In O5, it is seen a direct stile, a succinct expression by the enunciator of the information, strictly necessary. Moreover, it is integrated at macro-structural level. It should be noted as in O3, O4 and O5, Elena is written with "H". This is a clue to identify the processes of adaptation applied to the fixed texts of magic. It is more common in our environment to write Helena and not Elena.

Anyway, the analysis of several inscriptions about the same sortilege has us allowed evidencing how there is social traffic just by changing the arrangement of information, as to its structure and its contents. Although magic operates with fixed texts, however they can not escape to the social change by cultural influence. It is manifested itself in every actualization of the text by the ritual or ceremonial. Within these social changes, they are included the changes of intention and of purpose as is shown in O4.

Luis Carlos Molina Acevedo

The Symbolism of Sortilege

According to the classification of symbols by Gilbert Durand (1981), it is found in the sortilege of invocation to Santa Elena, an application of symbolism in its ceremonial. This can be interpreted semantically from the perspective referred by this author.

It is had in O1, O2 and O3 that the ceremonial remains unchanged in the use of symbols. There are three nails. Semantically, they represent the three schemes of symbols established by Gilbert Durand. The first nail given to Constantine acquires well a semantic isomorphism, corresponding to diairetic and vertical scheme symbolized by the sceptre and the sword. This nail will now be the sword. It will help to Constantine to win his battles, just as in the ceremonial, it serves to fight against spirits enemies.

The second nail is thrown to sea and therefore, this takes a semantic isomorphism. This characteristic puts it in the scheme of descent and internalizing, symbolized by the cup and its symbolic components, in this case the sea. In the ritual, it is the water where the content is the continent of one of the nails. To that extent, the magician, to achieve an end, manages to descend symbolically to the night. Now, he is able to obtain help from the spirit invoked, which is located in the Nocturnal Regimen.

The third nail to be nailed to the absent, as already gone, it acquires a semantic isomorphism. This feature places it in the rhythmic scheme that is a euphemism of time and its relentless course. Therefore, the time can enter into the rotation of the wheel and return as in the myth of the eternal return. And with it, it is returned the loved being.

Now, these three schemes are located in the Nocturnal Regime by a process of euphemism of Diurnal Regime. The ceremonial centre is the medal of Santa Elena, the Mother Goddess as euphemism of Father God. It is she who governs this symbolic ordering of the world. The water used here includes a semantic isomorphism to match the nocturnal water, to the female water par excellence: menstruation. But, this game of euphemisms is only intended to prevent the passage of time, breaking the course toward death. It is an attempt to operate the cyclical to do return the past, that past where one was happy with the loved one. The ritual, here, is presented as a time machine. If you can travel in time, you can control the courses taken by it. You can avoid the loss of a loved one. The ritual is presented as the safe way to travel back in time, without causing dislocations in it. This game of euphemisms explains why this sortilege is of greater use among women.

This symbolism is projected beyond reflexology. It reaches the field of culture where it is possible to identify a classification tools as well:

1. The first nail is placed in the category of useful forceful and percussive.

2. The second nail, by isomorphism, corresponds to content, found in the continents or containers (water).

3. The third nail is the return and, in turn, by a semantic isomorphism, imposes a technical extension of the wheel as transport. This moves away both, time and things. But, transport can also return them.

The first two nails are associated with excavation techniques, searching into the deep, in the beyond what was lost. It is needed to get it back to fill the underlying void in the ordering of world, in searching of the missing link. The ceremonial constitutes a double search in those two instances able to hide things: time and space. But, it really is only a two-dimensional search. It shows how by the semantics of symbols, it is possible to link each cultural practice of man to a universal cultural practice. The man in every act updates the universal culture.

In the same way as is operated for passing to the cultural, one can also pass to the social. It shows how in the ceremonial there also is a functional three-partition in social class. There is the presence of the sacerdotal class represented by Santa Elena to who is assigned the role of mystery, of link with the beyond. She ceases to belong to Diurnal Regime where she is placed by the religion, to pass to govern the Nocturnal Regime by a reversing in the value assigned to the terms of the antithesis: Father God/Mother Goddess. The warrior class is represented by the son of the Holy woman, by sailors who are the conquerors par excellence. They play a controversial role of constant challenges to the existing. And the producer class is represented by the portrait of the person who is looking to attract, who will not be a winner but a loser. To that extent, the user becomes a winner by a reversal of roles.

Finally, it should be noted how all this is true in O4 where it seems as if everything had been changed, but really only has a movement of isomorphism of symbols retaining the same semantics base. First, the candle ends still linked semantically to the little golden nails, because the ends are incomplete candles. Like what are the nails. They are not nails, but little nails. Instead of gold colour, the candle ends are amber, with colour of wax. Social classes are also present. The indigent represents the producer class, to that extent, he can give. The sacerdotal class is represented by the church, whose product (candle end) occupies the centre of the ceremonial. It is the mysterious, because there the three nails are also placed forming cross. And the warrior class, the controversial, is represented by the miser who conquers his treasures only in open struggle against others, always earrings take his possessions. It is the obsessive conqueror.

As for the symbolic schemes, it should be noted as the ceremonial centre is the product of Holy Mother Church, not the priest who is solicited the candle end. It corresponds to diairetic and vertical scheme by euphemism of the Diurnal Regimen of religion to settle on the Nocturnal Regimen where once again, the Mother Goddess is the centre of the ritual. The first candle end is placed on the left of the magician, that is, it is really to the right, taking as the axis of ordering of the world the image of Santa Elena. It fulfils the times of Scheme of descent and internalization. What is at right, it is considered some good. It satisfies the dominant reflex of nutrition. While the third candle end, by its position in the organization, acquires an isomorphism of something sinister. It is able to reverse things and dissolve the duration to return as well to chaos. It is propitiated as well a new beginning where rediscovering to the beloved absent.

In this ceremonial representation, it is less obvious the transition to the cultural, however, it is implicit and it is possible to recover it from a semantic projection of the macro-structure as well: it has that the second candle end, by coming from the Church, becomes the instrument of defence against spirits enemy. It acquires a semantic isomorphism of the percussive and forceful. The first candle end, by being of the producer class, is located on the scheme of descent according to the dominant reflex of nutrition, which is located in the belly. It acquires a semantic isomorphism of the continent and containers. The third candle end, by being at the left of the symbolic ordering of the world, takes a semantic isomorphism of transport as a technical extension of the wheel. It has the capacity to reverse the course of things. Thus, the search two-dimensional of magician is evidenced to find the desired beloved, to whom time and space are hidden.

Luis Carlos Molina Acevedo

Structure of Magical Discourse

The analysis performed about the inscriptions of magical discourse, allows advance to a higher level of abstraction. Therefore, it is necessary to classify and define categories and relations of the magical discourse as a whole, not its types by separated. The categories are identified in accordance to functions of them. To this extent, we can say that magical discourse is constituted globally by the following categories:

1. Instruction

2. Recipe

3. Petition

In the formulation of these categories, the order is only by enumerating and not structural. It has been found that they can occupy different positions in the magical discourse.

The **instruction** contains all the indications for the magician about the magical act. It explains each of the steps to achieve the intended effectiveness. The instruction corresponds with the category Action Plan. But this, it is not defined and charted by the agent or magician. It is previously set by the discourse itself. There is an enunciator of the discourse. He previously defined these instructions.

The category **recipe** is referred to the content. It relates to the preparation and combination of natural elements for implementing the ritual. In the magical-recipe is listed the follows:

a. What are the components required

b. Under what conditions should be collected and combined

c. What is the procedure required to be followed in its application

The category **petition** is represented by the texts to be recited. Of them, it is not important the reception of its semantic content, but its illocutionary force defined as a power to operate over things. So, the user hears only the murmur of who recites the text. That has done, these texts are called "secrets". They should only be known by magicians. In this feature is based the hypothesis proposed in this study: there is a third function of language, unspecified in specialized studies. This third function of language is directed to realize that deep-rooted belief in magic. With this, one can operate with language over things. It is a universal mental structure. It is evident not only in the ritual, but also in the doctrinal body of religions. This third function of language allows to the priest or minister or whoever is the representative of God on earth to forgive sins. By the power to operate with language in the world, it can be converted or transfigured the bread and wine into the body and blood of Jesus Christ. This is not only applicable to the West, it is also for all cultures, where the divine or not divine, can operate over the materiality of the world through the verb.

As for the functions performed by these categories within the structure of the magical discourse, we have:

1. The category **instruction** has essentially an educational function, instructive

2. The category **recipe** has an instrumental function for handling elements

3. The category **petition** has a function of enunciation. It concentrates the force of enunciation. It acts directly on the world without the will of the listener.

These functions can also be called of planning, of executing and of reinforcing, respectively. Thus, from the classification of the magical discourse, one can enter to establish the typology of the same as well:

1. The **magical-orison** will put its emphasis on the categories of instruction and petition

2. The **magical-secret** will focus on three categories, which constitutes the magical discourse par excellence: instruction, magical-recipe, and petition

3. The **magical-recipe** will focus on categories of instruction and recipe.

4. The **magical-formula** will emphasize on the categories of recipe and petition, with category of instruction as a presupposition.

Luis Carlos Molina Acevedo

Bibliography

ANAYA CAMPBELL, Eduardo. Secretos y recetas. Colombia, s.p.i. 203p.

AUSTIN, John, L. Cómo hacer cosas con palabras. Barcelona: Paidós, 1982. 217p.

BANDLER, Richard y GRINDER, John. La estructura de la magia I: lenguaje y terapia; trad.
Elena Olivos, Ataliva Amengual y Francisco Huneeus. Santiago de Chile: Cuatro Vientos, 1980. 226p.

BARYLKO, Jaime. La cábala: Orígenes, evolución y contenidos. Buenos Aires: Congreso Judío Latinoamericano, 1977. 47p.

BERGUA, Juan B., Trad. El libro de los muertos. Madrid: Gráficas Senén Marín, 1962. 554p.

BESSY, Maurice. Historia en 1000 imágenes de la magia, trad. Margarita García. Barcelona: Luis de Caralt, 1963. 367p.

BORIESSON, Maurice. La magia: sus grandes ritos y su historia, trad. Jesús Ruiz. Barcelona: Luis de Caralt, 1962. 411p.

181

BRUNO, Giordano, 1548-1600. Mundo, magia y memoria: Selección de textos / Giordano Bruno; ed. de Ignacio Gómez de Liaño. 2ed. Madrid: Taurus, 1982. 398p.

BUTLER, Alban. Vida de santos de Butler II. México: Collier's International - John W. Clute, 1985. 701p.

CARRASQUILLA, Tomás. La marquesa de Yolombó, ed. Crítica de Kurt L. Levy. Bogotá: Instituto Caro y Cuervo, 1974. 630p.

CARVAJAL P., Jorge. Bases científicas de una medicina bioenergética. En: Memorias Simposio de medicina tradicional 200 años. Medellín: Tipografía Suárez, 1988. 215p.

_____ Hacia una edad sin tiempo. En: La bioenergética hoy, I Congreso de medicina bioenergética. Medellín: Papeles y Copias, 1986. 62p.

_____ Resonacia Mórfica. Ibidid.

CASSIRER, Ernest. Antropología filosófica, trad. Eugenio Imaz. 12ed. México: Fondo de Cultura Económica, 1987. 335p.

CASTIGLIONI, Arturo. Encantamiento y magia, trad. Guillermo Pérez Enciso. México: Fondo de Cultura Económica, 1947. 426p.

CAZENEUVE, Jean. Sociología del rito, trad. José Castelló. Buenos Aires: Amorrortu, 1971. 279p.

CORONA MÍSTICA. México: Juana de Arco, s.f. 159p.

DUCROT, Oswald. Polifonía y argumentación: Conferencias del Seminario Teoría de la Argumentación y análisis del discurso, trad. Ana Beatriz Campo y Emma Rodríguez C. Cali: Universidad del Valle, 1990. 190p.

DURAND, Gilbert. Las estructuras antropológicas de lo imaginario, trad. Mauro Armiño. Madrid: Taurus, 1981. 453p.

ELIADE, Mircea. El mito del eterno retorno, trad. Ricardo Anaya. Barcelona: Planeta - Agostini, 1985. 164p.

ELIPHAS LEVI, Seud. de Constant Alphonse Louis, 1816-1875. Historia de la magia: Resumen de sus procedimientos, ritos y misterios; trad. Enrique Barea. Madrid: El Adelantado de Segovia, 1922. 491p.

_____ La clave de los misterios. Guayaquil: Ariel, 1975. 174p.

ESCOBAR URIBE, Arturo. Rezanderos y ayudados. 2ed. Bogotá: Gráficas Venus, 1967. 261p.

FOUCAULT, Michel. La palabras y las cosas, trad. Elsa Cecilia Frost. México: Siglo Veintiuno, 1984. 385p.

FREEDLAND, Nat. La explosión del ocultismo, trad. René Cárdenas Barrios. México: Diana, 1973. 285p.

FUENTES, Carlos. La nueva novela hispanoamericana. 3ed. México: Joaquín Mortiz, 1972. 98p.

GARCIA - FONT, Jean. Magia - Brujería - Demonología. Barcelona: Glosa, 1977. 282p.

HUBBARD, L. Ronald. Dianética: la evolución de una ciencia. 2ed. México: Dazet, 1977.   86p.

JUNG, Carl G. Realidad del alma, trad. Fernando Vela y Felipe Jiménez de Asús. 4ed.   Buenos   Aires: Losada, 1968. 171p.

_____ Símbolos de transformación. 2ed. Buenos Aires: Paidós, 1962. 441p.

KARDONNE, G. El verdadero significado de la ceniza del cigarrillo. Medellín: Antorcha,   1975. 99p.

LACAN, Jacques. Escritos I. México: Siglo Veintiuno, 1981. 374p.

LARA, Karen. Recetario de magia blanca. 2ed. Bogotá: Lito - Artes Guerrero's, 1989 171p.

LEVI - STRAUSS, Claude. Antropología estructural, trad. Eliseo Verón. 2ed. Buenos   Aires: Universitaria, 1969. 371p.

_____ Mitológicas: Lo crudo y lo cocido I. México: Fondo de Cultura Económica.   1986. 395p.

LIBRO DE San Cipriano: Tesoro de las ciencias ocultas. México, s.e.f. 159p. (Biblioteca   Ciencias Ocultas).

LIBRO INFERNAL: El tesoro de la ciencias ocultas. México: Saturno, 1969. 459p.

LÓPEZ TORO, José Hernán. La homeopatía: Medicina energética. En: La bioenergética   hay,   I

Congreso de medicina bioenergética. Medellín: Papeles y Copias, 1986. 62p.

MOLINA ACEVEDO, Luis Carlos. Imaginaria de la exageración. En: Cuadernos académicos "Quirama". Medellín, No. 9 (may., 1990), p. 39-56.

MOORNE, Doctor. Enchiridiones Grimorios y pantáculos. 9ed. México: La Campana, s.f. 232p.

PAPUS, Seud. de Dr. G. Encause. Tratado de magia práctica. 9ed. Buenos Aires: Kier, 1988. 532p.

_____ Tratado elemental de ciencia oculta. 2ed. Buenos Aires: Kier, 1978. 249p.

PAYAN DE LA ROCHE, Julio César. Bases generales de la Bio-Cibernética aplicadas a la terapia neuronal y a la homotoxicología. En: La bioenergética hoy, I Congreso de medicina bioenergética. Medellín: Papeles y Copias, 1986, 62p.

ROTUNDO, Emiro. Introducción a la teoría general de los sistemas. 5ed. Caracas: Universidad Central de Venezuela, Facultad de Ciencias Económicas y sociales, División de Publicaciones, 1985. 96p.

RYZL, Milan. Cómo potenciar sus poderes paranormales, trad. Jordi Fibla. Bogotá: Lerner, 1991. 188p.

SALVERTE, Eusebio, 1771-1839. Las ciencias ocultas: Ensayo sobre la magia, los prodigios y milagros; trad. D.F.J. Orellana. Barcelona: Imprenta y librería de Salvador Moreno, 1865. 446p.

SANTA CRUZ de Caravaca, la: Tesoro de oraciones. México, s.e.f. 144p.

SEARLE, John. Actos de habla, trad. Luis M. Valdés Villanueva. Madrid: Cátedra, 1990. 201p.

SÉJOURNÉ, Laurette. Supervivencias de un mundo mágico. México: Fondo de Cultura Económica, 1953. 116p.

SCARDUELLI, Pietro. Dioses, espíritus, ancestros: Elementos para la comprensión de sistemas rituales; trad. Guillermo Pérez Enciso. México: Fondo de Cultura Económica, 1988. 195p.

SCHOPENHAUER, A. Las ciencias ocultas. Buenos Aires: Kier, 1955. 181p.

TONDRIAU, Julie. Los enigmas del ocultismo, trad. Julio Moreno Bernardo. Madrid: Daimón, 1966. 403p. (Colección Grandes Enigmas V. 13).

VAN DIJK, Teun A. Texto y contexto, trad. Juan Domingo Moyano. 3ed. Madrid: Cátedra, 1988. 357p.

Images

Ilustración 1

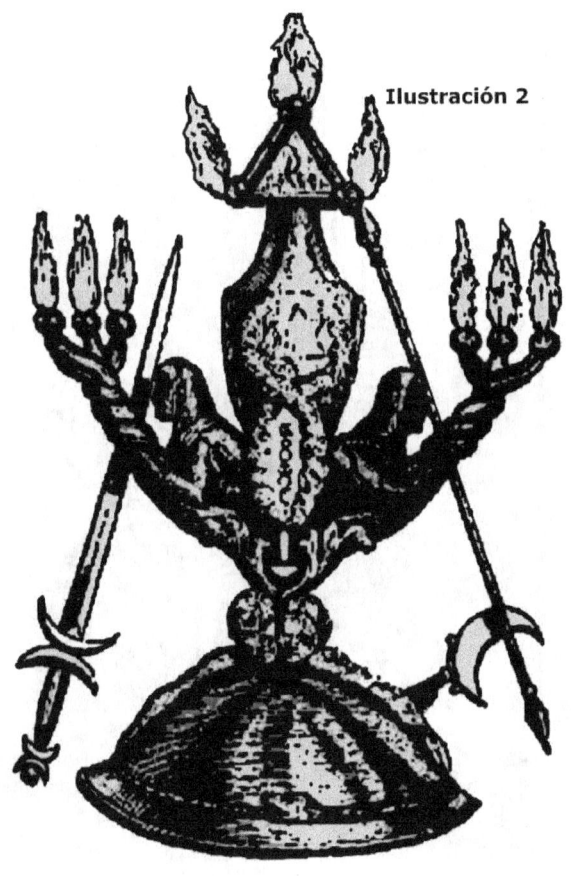

Ilustración 2

Luis Carlos Molina Acevedo

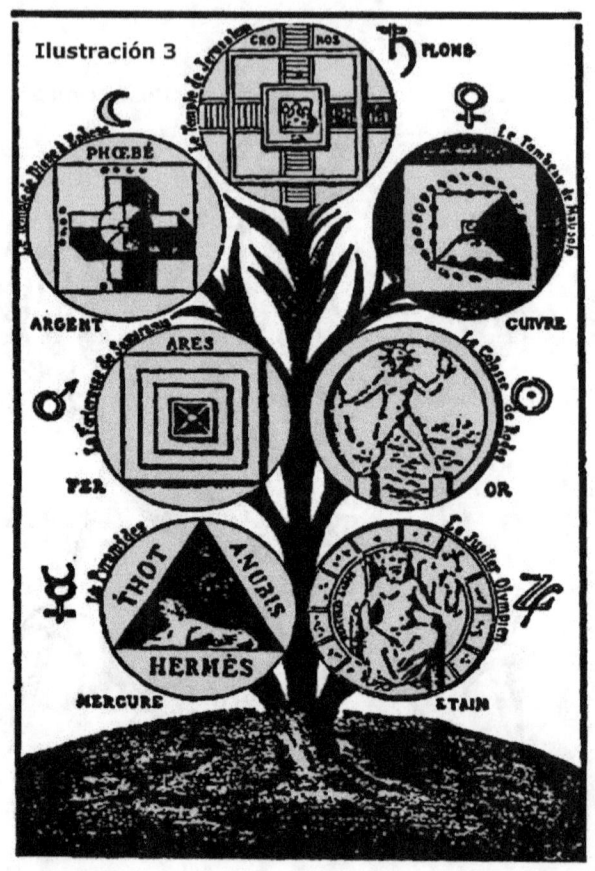

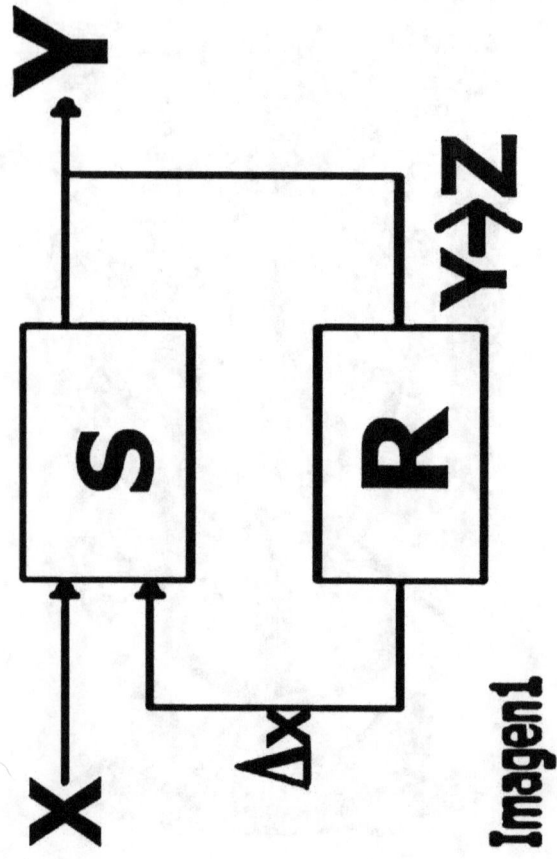

Imagen1

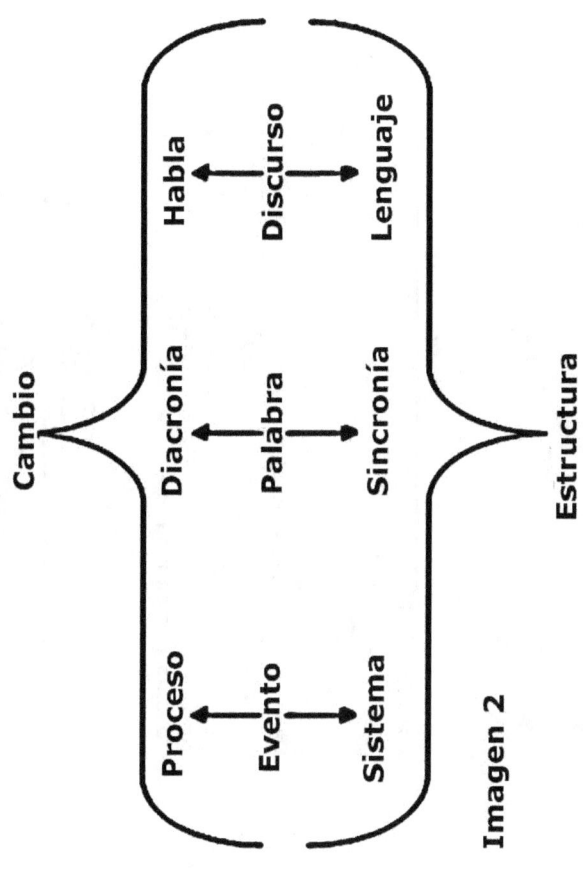

Imagen 2

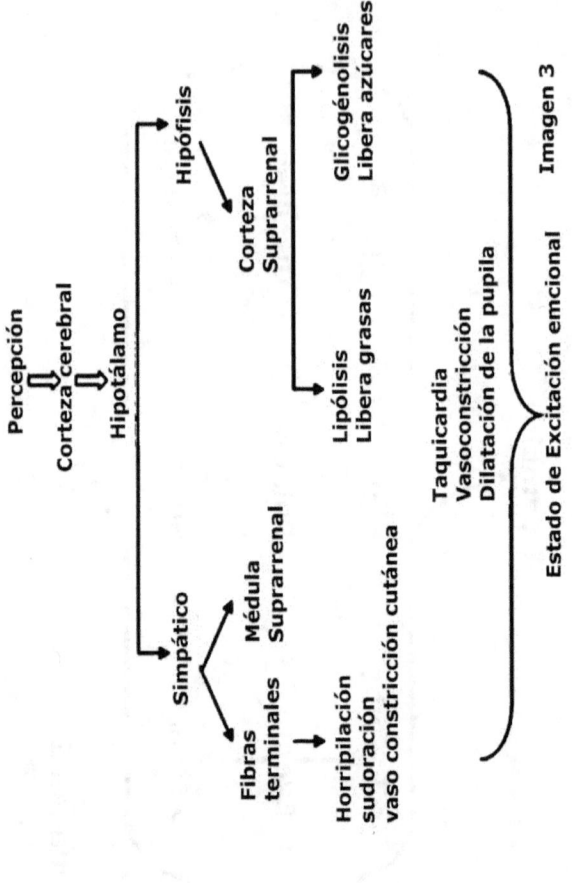

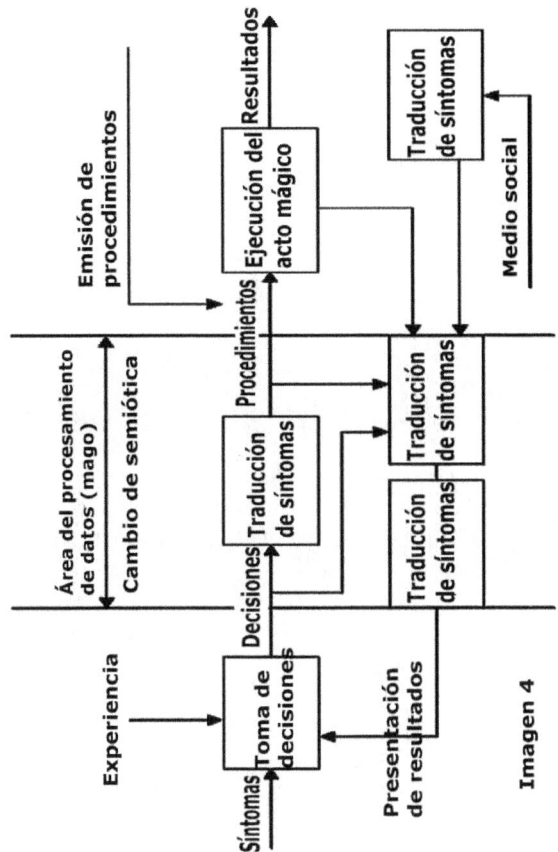

Imagen 4

Luis Carlos Molina Acevedo